157
Advances in Polymer Science

Editorial Board:
A. Abe · A.-C. Albertsson · H.-J. Cantow
K. Dušek S. Edwards · H. Höcker
J. F. Joanny · H.-H. Kausch · K.-S. Lee
J. E. McGrath · L. Monnerie · S. I. Stupp
U. W. Suter · G. Wegner · R. J. Young

Springer
Berlin
Heidelberg
New York
Barcelona
Hong Kong
London
Milan
Paris
Tokyo

Degradable Aliphatic Polyesters

Volume Editor: A.-C. Albertsson

With contributions by
A.-C. Albertsson U. Edlund, M. Hakkarainen,
S. Karlsson, Y. Liu, E. Ranucci, M. Ryner,
M. Söderqvist Lindblad, K. M. Stridsberg,
I. K. Varma

 Springer

This series presents critical reviews of the present and future trends in polymer and biopolymer science including chemistry, physical chemistry, physics and materials science. It is addressed to all scientists at universities and in industry who wish to keep abreast of advances in the topics covered.

As a rule, contributions are specially commissioned. The editors and publishers will, however, always be pleased to receive suggestions and supplementary information. Papers are accepted for "Advances in Polymer Science" in English.

In references Advances in Polymer Science is abbreviated Adv Polym Sci and is cited as a journal.

Springer APS home page: http://link.springer.de/series/aps/ or
http://link.springer-ny.com/series/aps/
Springer-Verlag home page: http://www.springer.de

ISSN 0065-3195
ISBN 3-540-42249-8
Springer-Verlag Berlin Heidelberg New York

Library of Congress Catalog Card Number 61642

This work is subject to copyright. All rights are reserved, whether the whole or part of the material is concerned, specifically the rights of translation, reprinting, re-use of illustrations, recitation, broadcasting, reproduction on microfilms or in other ways, and storage in data banks. Duplication of this publication or parts thereof is only permitted under the provisions of the German Copyright Law of September 9, 1965, in its current version, and permission for use must always be obtained from Springer-Verlag. Violations are liable for prosecution under the German Copyright Law.

Springer-Verlag Berlin Heidelberg New York
a member of BertelsmannSpringer Science+Business Media GmbH

http://www.springer.de

© Springer-Verlag Berlin Heidelberg 2002
Printed in Germany

The use of registered names, trademarks, etc. in this publication does not imply, even in the absence of a specific statement, that such names are exempt from the relevant protective laws and regulations and therefore free for general use.

Typesetting: Data conversion by MEDIO, Berlin
Cover: MEDIO, Berlin
Printed on acid-free paper   SPIN: 10741878   02/3020hu - 5 4 3 2 1 0

Editorial Board

Prof. Akihiro Abe
Department of Industrial Chemistry
Tokyo Institute of Polytechnics
1583 Iiyama, Atsugi-shi 243-02, Japan
E-mail: aabe@chem.t-kougei.ac.jp

Prof. Ann-Christine Albertsson
Department of Polymer Technology
The Royal Institute of Technolgy
S-10044 Stockholm, Sweden
E-mail: aila@polymer.kth.se

Prof. Hans-Joachim Cantow
Freiburger Materialforschungszentrum
Stefan Meier-Str. 31A
D-79104 Freiburg i. Br., FRG
E-mail: cantow@fmf.uni-freiburg.de

Prof. Karel Dušek
Institute of Macromolecular Chemistry, Czech
Academy of Sciences of the Czech Republic
Heyrovský Sq. 2
16206 Prague 6, Czech Republic
E-mail: dusek@imc.cas.cz

Prof. Sam Edwards
Department of Physics
Cavendish Laboratory
University of Cambridge
Madingley Road
Cambridge CB3 OHE, UK
E-mail: sfe11@phy.cam.ac.uk

Prof. Hartwig Höcker
Lehrstuhl für Textilchemie
und Makromolekulare Chemie
RWTH Aachen
Veltmanplatz 8
D-52062 Aachen, FRG
E-mail: hoecker@dwi.rwth-aachen.de

Prof. Jean-François Joanny
Institute Charles Sadron
6, rue Boussingault
F-67083 Strasbourg Cedex, France
E-mail: joanny@europe.u-strasbg.fr

Prof. Hans-Henning Kausch
c/o IGC I, Lab. of Polyelectrolytes
and Biomacromolecules
EPFL-Ecublens
CH-1015 Lausanne, Switzerland
E-mail: kausch.cully@bluewin.ch

Prof. Kwang-Sup Lee
Department of Polymer Science & Engineering
Hannam University
133 Ojung-Dong
Teajon 300-791, Korea
E-mail: kslee@mail.hannam.ac.kr

Prof. James E. McGrath
Polymer Materials and Interfaces Laboratories
Virginia Polytechnic and State University
2111 Hahn Hall
Blacksbourg
Virginia 24061-0344, USA
E-mail: jmcgrath@chemserver.chem.vt.edu

Prof. Lucien Monnerie
École Supérieure de Physique et de Chimie
Industrielles
Laboratoire de Physico-Chimie
Structurale et Macromoléculaire
10, rue Vauquelin
75231 Paris Cedex 05, France
E-mail: lucien.monnerie@espci.fr

Prof. Samuel I. Stupp
Department of Measurement Materials Science
and Engineering
Northwestern University
2225 North Campus Drive
Evanston, IL 60208-3113, USA
E-mail: s-stupp@nwu.edu

Prof. Ulrich W. Suter
Department of Materials
Institute of Polymers
ETZ,CNB E92
CH-8092 Zürich, Switzerland
E-mail: suter@ifp.mat.ethz.ch

Prof. Gerhard Wegner
Max-Planck-Institut für Polymerforschung
Ackermannweg 10
Postfach 3148
D-55128 Mainz, FRG
E-mail: wegner@mpip-mainz.mpg.de

Prof. Robert J. Young
Manchester Materials Science Centre
University of Manchester and UMIST
Grosvenor Street
Manchester M1 7HS, UK
E-mail: robert.young@umist.ac.uk

Preface

Aliphatic polyesters are, together with polycarbonates, polyanhydrides, and poly(amino acids), the most well-known synthetic hydrolyzable polymers. They are often prone to degradation but are at the same time usually not good enough for technical applications. A renewed interest in aliphatic polyesters has resulted in developing new materials important in the biomedical and ecological fields.

In our first chapter, we summarize the synthesis of aliphatic polyesters. This includes homopolyesters, random, block, graft, and star- and hyper-branched polyesters. Mainly materials such as PLA and PCL homopolymers have so far been used in most applications. There are, however, many others monomers which one can use as homopolymers or in copolymerization with lactide and caprolactone. Different molecular stuctures give a wider range of physical properties as well as the possibility of regulating the degradation rate.

In the second chapter, we try to emphasize the possibilities of producing tailor-made polymers with predicted properties. By using different types of initiators and catalysts, ring-opening polymerization of lactones and lactides provides macromolecules with advanced molecular architecture – a careful selection of appropriate conditions is crucial. The purpose of this chapter is also to describe the mechanisms and typical kinetic features.

One of the possible applications for degradable polymers is drug delivery. In the third chapter, we have chosen to describe the use of different degradable polymers in this application. This review comprises degradable polymers for use in controlled drug delivery with emphasis on the preparation, applications, and biocompatibility.

In any application, the use of degradable polymers is dependent on the knowledge of the interaction of these materials with the environment. The type and the number of small molecules evolving from the materials govern this interaction. The fourth chapter is therefore devoted to degradation products and chromatographic measurements. The amounts of degradation products, as well the kind of products from one polymer, change with time and with the environment. The chosen test method also determines the degradation rate.

In the future, new degradable polymers should be able to participate in the metabolism of nature. In the last chapter we give some examples of novel polymers with inherent environmentally favorable properties such as renewability and degradability and a series of interesting monomers found in the metabolisms and cycles of nature. The results reported demonstrate renewable polymers with a variety of physical and mechanical properties.

Finally, I wish to thank all my co-workers who made this work possible.

Stockholm, Summer 2001   Ann-Christine Albertsson

Advances in Polymer Science
Now Also Available Electronically

For all customers with a standing order for Advances in Polymer Science we offer the electronic form via LINK free of charge. Please contact your librarian who can receive a password for free access to the full articles. By registration at:

http://link.springer.de/series/aps/reg_form.htm

If you do not have a standing order you can nevertheless browse through the table of contents of the volumes and the abstracts of each article at:

http://link.springer.de/series/aps/
http://link.springer-ny.com/series/aps/

There you will find also information about the

- Editorial Bord
- Aims and Scope
- Instructions for Authors

Contents

Aliphatic Polyesters: Synthesis, Properties and Applications
A.-C. Albertsson, I. K. Varma 1

Controlled Ring-Opening Polymerization: Polymers
with designed Macromolecular Architecture
K. M. Stridsberg, M. Ryner, A.-C. Albertsson 41

Degradable Polymer Microspheres for Controlled Drug Delivery
U. Edlund, A.-C. Albertsson 67

Aliphatic Polyesters: Abiotic and Biotic Degradation
and Degradation Products
M. Hakkarainen 113

Polymers from Renewable Resources
M. Söderqvist Lindblad, Y. Liu, A.-C. Albertsson, E. Ranucci, S. Karlsson ... 139

Author Index Volumes 101–157 163

Subject Index 177

Aliphatic Polyesters: Synthesis, Properties and Applications

Ann-Christine Albertsson[1], Indra K. Varma[2]

[1] Department of Polymer Technology, The Royal Institute of Technology, 10044 Stockholm, Sweden
 e-mail: aila@polymer.kth.se
[2] Centre for Polymer Science and Engineering, Indian Institute of Technology, Hauz Khas, New Delhi-110016, India
 e-mail: ikvarma@hotmail.com

Abstract. The synthesis of aliphatic polyesters by polycondensation and ring-opening polymerization is reviewed. This includes homopolyesters, random, block, graft, star, and hyper-branched (co)polyesters. Recent progress in the synthesis of high molecular weight aliphatic polyesters is described. Specific properties of these polymers are also given. The biomedical and ecological applications of these biodegradable polymers show their technological importance and relevance.

Keywords. Polycondensation, Ring-opening polymerization, Lactones, Biocompatible, Biodegradable, Bioresorbable, Hyper-branched polymers

1	Introduction	2
2	Synthesis	4
2.1	Polycondensation	4
2.1.1	Monomers	4
2.1.2	Polymerization	4
2.2	Ring-Opening Polymerization	7
2.2.1	Monomers	7
2.2.2	Polymerization	10
2.2.2.1	Free Radical Polymerization	11
2.2.2.2	Anionic Polymerization	12
2.2.2.3	Carbocationic Polymerization	15
2.2.2.4	Coordinative Ring-Opening Polymerization of Lactones	16
2.3	Functional Polyesters	23
2.4	Hyper-Branched Polyesters	23
3	Physical Properties	26
4	Degradation	29
4.1	Homopolymers	30
4.2	Copolyesters	32

| 5 | Applications | 34 |

References 35

Abbreviations

Mn	number-average molecular weight
Mw	weight-average molecular weight
DPn	number-average degree of polymerization
$[M]_o$	initial monomer concentration
$[I]_o$	initial concentration of initiator
THF	tetrahydrofuran
ROP	Ring-opening polymerization
β-PL	β-propiolactone
3-HB	3-hydroxybutyrate
γ-BL	γ-butyrolactone
HA	hydroxyalkanoates
LA	lactide
ε-CL	ε-caprolactone
δ-VL	δ-valerolactone
PLA	polylactide
PLLA	poly(L-lactide)
PDLA	poly(D-lactide)
PDLLA	poly(D,L-lactide)
bis-MPA	2,2-bis(hydroxymethyl)propionic acid
HEMA	2-hydroxyethyl methacrylate
DXO	1,5-dioxepan-2-one
DDXO	1,5,8,12-tetraoxacyclotetradecane-2,9-dione
THP	tetrahydro-4*H*-pyran-4-one
PET	poly(ethylene terephthalate)
PGA	poly(glycolic acid)

1
Introduction

Aliphatic polyesters are among the most used biodegradable polymers in medical applications and have been extensively investigated in the past. Ester linkages are frequently encountered in nature and hence it is expected that synthetic polymers containing such linkages and an appropriate structure would be environmentally degradable. For example, shellac, a naturally occurring aliphatic polyester known from time immemorial, is biodegradable. Its applications are described in the well-known Indian epic Mahabharatha believed to be written around 3000 BC. This biodegradable resin, a mixture of aliphatic polyhydroxy acids present as lactones and polyesters, was used extensively in protective coatings. It was the success of shellac that led Bakeland to develop phenolic resins as its substitute and hence lay the foundation for the synthetic polymer industry at the turn of the last century. It is noteworthy that at the beginning of the third millennium we are once again looking at nature as a model for developing bioadaptable and environmentally adaptable polymers [1]. The superior performance of natural polymers can be attributed to their multifunctionality. Separate components have more than one function, as is evident in wood where a synergistic combination of fiber (cellulose), hemicellulose, and lignin (matrix) leads to a strong, light-weight material.

Carothers' pioneering studies were also based on aliphatic polyesters and then culminated in laying the foundations for condensation and step-growth polymerization and in establishing a relationship between molar mass and extent of reaction and the stoichiometric imbalance of functional groups. Fundamental studies relating structure to properties were carried out using these polymers.

The aliphatic polyesters, developed initially, were of low molecular mass with poor mechanical properties. The dihydroxy-terminated poly(alkylene alkanoates) found applications in the production of polyurethane [2] or as plasticizers in PVC [3]. In the 1960s poly(L-lactide), PLLA, was proposed as a biocompatible, biodegradable, and bioresorbable material (a material that degrades and is further resorbed *in vivo*) [4]. Homopolymers and copolymers of lactide and glycolide form the basis of surgical sutures, drug delivery devices, and other body implants [5–8]. The most widely used absorbable sutures are Dexon, a multifilament PGA, and Vicryl, a copolymer with composition poly[L-LA (8%)-co-GA (92%)] and poly(*p*-dioxanone) PDS. In recent years, environmental concerns have led to a renewed interest in biodegradable polyesters as an alternative to commodity plastics [9, 10]. Poly(butylene succinate), poly(butylene succinate-adipate) copolymer, and poly(ethylene succinate) have successfully been prepared through condensation reactions of glycols with aliphatic dicarboxylic acids [11] under the trade mark of BIONOLLE. New grades having a long chain branch, have recently been developed for stretched blown bottles and expanded foams. Ring-opening polymerization (ROP) of lactones has yielded polyesters with very high molecular mass and good mechanical properties. Random and block copolyesters as well as blends have been investigated [12] with regard to the regulation of biodegradation and improvement in the mechanical properties.

Several review articles on biodegradable polymers and polyesters have appeared in the literature [12–22]. Extensive studies have been carried out by Albertsson and coworkers developing biodegradable polymers such as polyesters, polyanhydrides, polycarbonates, etc., and relating the structure and properties of aliphatic polyesters prepared by ROP and polycondensation techniques. In the present paper, the current status of aliphatic polyesters and copolyesters (block, random, and star-shaped), their synthesis and characterization, properties, degradation, and applications are described. Emphasis is placed primarily on aliphatic polyesters derived by condensation of diols with dicarboxylic acids (or their derivatives) or by the ROP of cyclic monoesters. Polyesters derived from cyclic diesters or microbial polyesters are beyond the scope of this review.

2
Synthesis

The traditional way of synthesizing polyesters has been by polycondensation using diols and a diacid (or an acid derivative), or from a hydroxy acid. This method suffers, however, from some major shortcomings. These are the need for (a) high temperature, (b) long reaction times, (c) removal of reaction by-products, and (d) a precise stoichiometric balance between reactive acid and hydroxy groups. Very high conversion is desirable to get polymer chains of sufficiently high molecular masses to provide useful mechanical properties in the final product. In spite of all precautions, a high degree of polymerization is very difficult to achieve by this method because of side-reactions and the volatilization of monomers, which leads to a stoichiometric imbalance of reactants.

The ROP of lactones, cyclic diesters (lactides and glycolides), and cyclic ketene acetals is an alternative method, which has been successfully employed to yield high molecular mass polymers under relatively mild conditions. This polyaddition reaction can be carried out with no or very limited side-reactions, and this makes it possible to control properties like molecular weight and molecular weight distribution (MWD). Both polycondensation and ROP have been reported in the literature for the synthesis of polyesters.

2.1
Polycondensation

2.1.1
Monomers

The monomers used in polycondensation reactions are generally petroleum-based, although some of these are also obtained from renewable resources. For example, the production of 1,3-propanediol by fermentation of glycerol was reported in 1881. *Clostridia* as well as *Enterobacteriaceae* can carry out this conversion [23]. Succinic acid can be produced by the fermentation of carbohydrates such as glucose, sucrose, maltose, or fructose using *Anaerobiospirillum succiniciproducens* [24]. Glucose fermentation using *Actinobacillus succinogenes* has been reported to give good yields of succinate [25].

2.1.2
Polymerization

High molecular weight random copolyesters of poly(ethylene terephthalate) (PET) with tetra-/hexa-/octa-/poly(ethylene glycol) prepared by polycondensation reactions have been reported by Varma et al. [26–32]. Such copolyesters are not biodegradable. Degradable block copolyesters of PET with poly(ethylene glycol) [33, 34] or poly(tetramethylene glycol) leave aromatic oligomers within the human body long after they have degraded. Studies on copolyesters containing optimal contents of aliphatic and aromatic moieties have shown that it is possible to combine both microbial degradability and favorable thermal and mechanical properties [35–37]. Fast degrading aliphatic block copolyesters (degrading within 10 days) have also been developed [38]. Biodegradable aliphatic polyesters trademarked as BIONOLLE are produced through the polycondensation reaction of 1,4-butanediol, succinic acid (#1000 series) and adipic acid (#3000 series) or ethylene glycol and succinic acid (#6000 series) [11]. Chain extension by a coupling reaction has been carried out to obtain high molecular weights. A semi-commercial plant with a capacity of 3,000 tons/annum has been constructed in 1993 for its production in Japan (Showa Co.).

Albertsson and Ljungquist [39] synthesized ^{14}C labeled poly(tetramethylene adipate) using adipic acid, tetramethylene glycol, and titanium tetraisopropoxide as a catalyst. The reaction was carried out in a nitrogen atmosphere for several hours at 190 °C and, when water evolution ceased, the temperature was lowered to 140–150 °C. Further reaction took place at reduced pressure (1 torr) at 165 °C for several hours until the contents became very viscous. The polymer was recovered by precipitating its acetone solution in methanol. Polycondensation of 3-hydroxyalkanoic acids and their derivatives has also been reported in the literature [40]. For example, polymerization of methyl (R)-3-hydroxybutyrate and its mixture with methyl (R)-3-hydroxyvalerate was carried out at atmospheric pressure using titanium(IV) isopropoxide at 140 °C for 2 h in a nitrogen atmosphere. The pressure was then reduced to 0.5 torr and the mixture was stirred for 3 h at this pressure and temperature, followed by 30 h at 10^{-3} torr. The Mw and Mn of these polyesters were in the range of 10,000–16,000. A dehydration reaction (a side-reaction) is a problem in such polycondensation reactions.

As mentioned earlier, it is often difficult to prepare high molecular weight polymers using the polycondensation method. Oligomeric products with a molecular weight in the range of a few thousands are easily obtained. These oligomers do not possess suitable thermal and mechanical properties for most practical applications. However, a chain extension reaction may be carried out with such oligomers. Thus, polyesters have been synthesized by polycondensation of octadecanedicarboxylic acid with caprolactone diol and the sequential chain extension reaction with sebacoyl chloride [41].

A degradable aliphatic thermoplastic elastic block copolymer, poly(ethylene glycol)/poly(ethylene succinate) was synthesized in a two-step process by polycondensation of dimethyl succinate (0.6 mol), ethylene glycol (0.9 mol), and poly(ethylene glycol) (0.05 mol) followed by chain extension using adipoyl chloride [42]. The polycondensation was done at 240 °C for several hours. Block copolyesters-poly[(ethylene succinate)-b-poly(tetramethylene glycol)] with Mn ranging from 25,000–45,000 have been prepared by polycondensation using titanium tetraisopropoxide as catalyst [43].

Recently, α,ω-bis-hydroxy-terminated poly(1,3-propylene succinate) has been chain-extended to yield high molecular weight poly(ester-carbonates) [44] using a bischloroformate route. Thus, using a molar ratio of 1,3-propanediol to succinic acid of 1.02, an oligomer having Mn of 2,200 and Mw of 3,000 was ob-

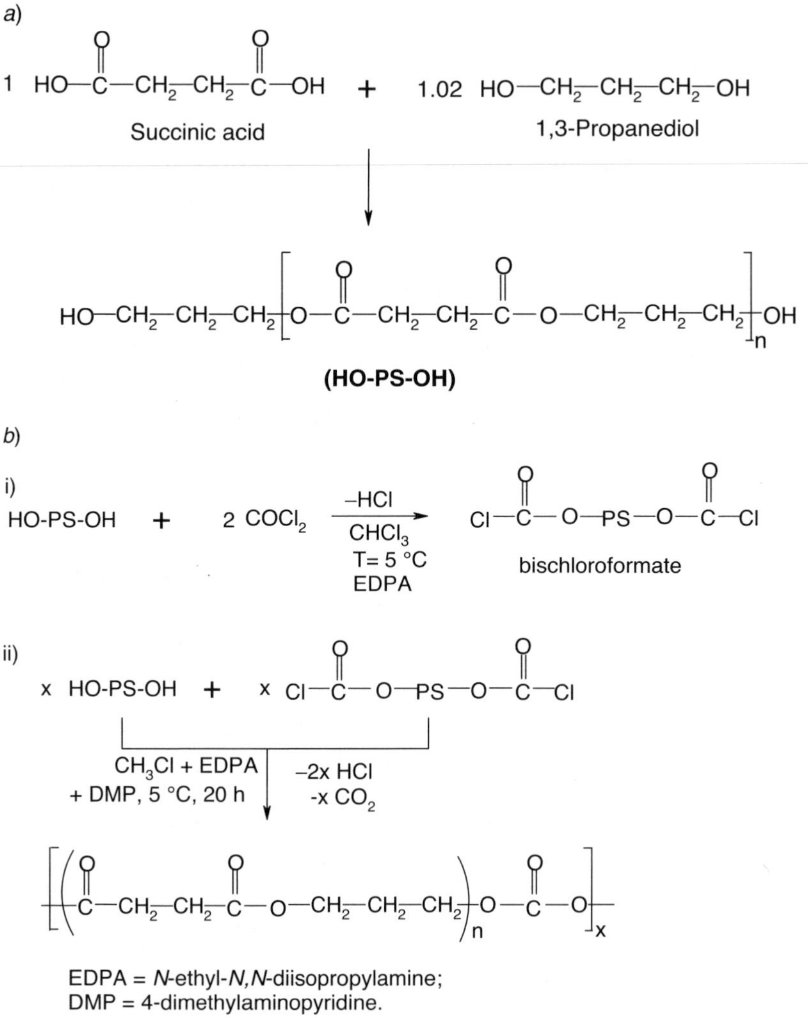

Scheme 1. *a)* Synthesis of dihydroxyterminated oligomeric propyl succinate. *b)* Chain extension reaction by the dichloroformate route

tained (Scheme 1, *a*). α,ω-Bischloroformate was then prepared by treating this oligomer with phosgene. Poly(ester-carbonate) (M_n=30,000 and M_w=48,000) was prepared by polycondensation with α,ω-bishydroxy terminated poly(1,3-propylene succinate) (Scheme 1, *b*). Poly(ester-urethane) could be prepared by chain extension with diisocyanates (Scheme 1, *c*).

Lipase-catalyzed polycondensation and transesterification reactions are the subjects of intensive research activities but polyesters of low molecular weight are obtained by this technique [45–52].

c)

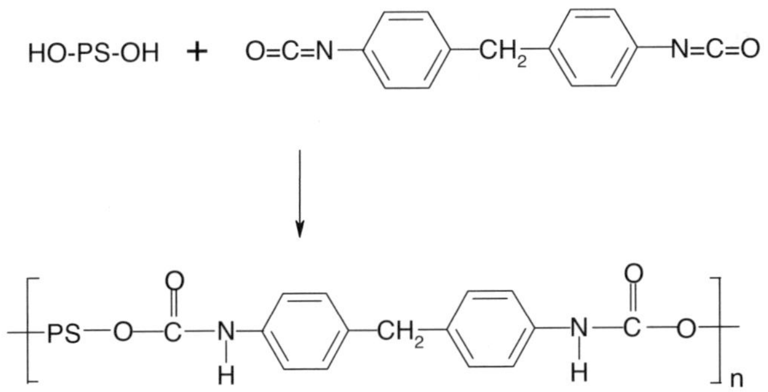

HO—PS—OH + O=C=N—⟨⟩—CH₂—⟨⟩—N=C=O

↓

[—PS—O—C(=O)—N(H)—⟨⟩—CH₂—⟨⟩—N(H)—C(=O)—O—]ₙ

Scheme 1. *c)* Preparation of poly(ester-urethane)

2.2
Ring-Opening Polymerization

2.2.1
Monomers

The monomers that have been used for the synthesis include glycolide, lactide, β-propiolactone, β-butyrolactone, γ-butyrolactone, δ-valerolactone, ε-caprolactone, 1,5-dioxepan-2-one, pivalolactone, 1,4-dioxane-2-one, 2-methylene-1,3-dioxolane, 2-methylene-1,3-dioxepane, etc. The structures of some of these monomers are given in Table 1.

Monomers like glycolide or lactide are prepared by heating the corresponding acids under controlled conditions [53]. For example, lactide is prepared by heating lactic acid at 120 °C until water ceases to distill. The temperature is then increased to 140 °C and the pressure is reduced to 10 torr. After heating for several hours at this temperature, the pressure is reduced further and the temperature increased until lactide begins to distill. Dilactide (3,6-dimethyl-1,4-dioxan-2,5-dione) contains two asymmetric centers and therefore exists as L-lactide, D-lactide, *meso*-lactide, and the racemic mixture D,L-lactide and gives polymers with different properties.

β-Lactones are generally prepared by reacting ketenes with carbonyl compounds or cyclization of salts of β-halo acids (Scheme 2)

The preparation of β-propiolactone involves bubbling an equimolar mixture of gaseous ketene and formaldehyde into a solution of β-PL containing a small amount of a complex catalyst composed of $AlCl_3$ and $ZnCl_2$. β-PL is a carcinogenic compound and has to be handled with care.

The synthesis of 1,5-dioxepan-2-one (DXO) was earlier reported from acrylonitrile and ethylene glycol [54]. Another approach was via ring closure of methyl 3-(2-hydroxyethoxy)propionate by an organometallic transesterification cata-

Table 1. Structure and designation of various lactones

Name	Structure	Designation
β-propiolactone		β-PL
γ-butyrolactone		γ-BL
β-butyrolactone		β-BL
δ-valerolactone		δ-VL
ε-caprolactone		ε-CL
1,5-dioxepan-2-one		DXO
R=H; glycolide		GA
R=CH$_3$; lactide		LA

lysts. However, a poor yield of DXO was obtained using this method and some polymer was formed as a side product. Mathiesen and Albertsson [55] investigated the synthetic route and characterized the monomer and its cyclic dimer 1,5,8,12-tetraoxacyclotetradecane-2,9-dione (DDXO). They attributed the poor yield of DXO to the formation of DDXO and oligomeric materials that complicate its isolation.

A new procedure (Scheme 3) using the Baeyer-Villiger oxidation of tetrahydro-4*H*-pyran-4-one (THP) was developed by Mathisen et al., which gave a better yield with no side-reactions, thereby facilitating the isolation of DXO [56].

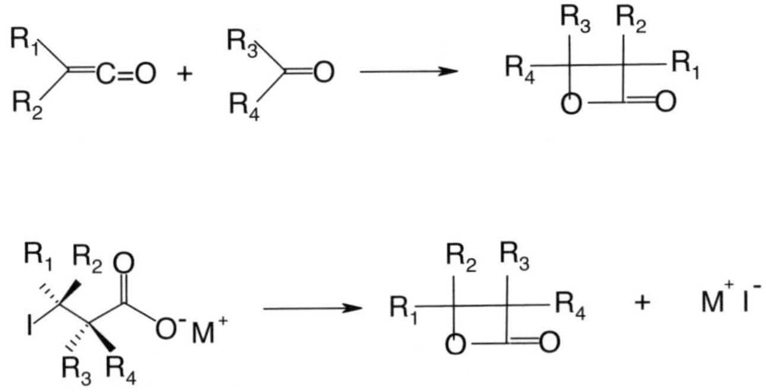

Scheme 2. Reaction sequence for the synthesis of β-lactones

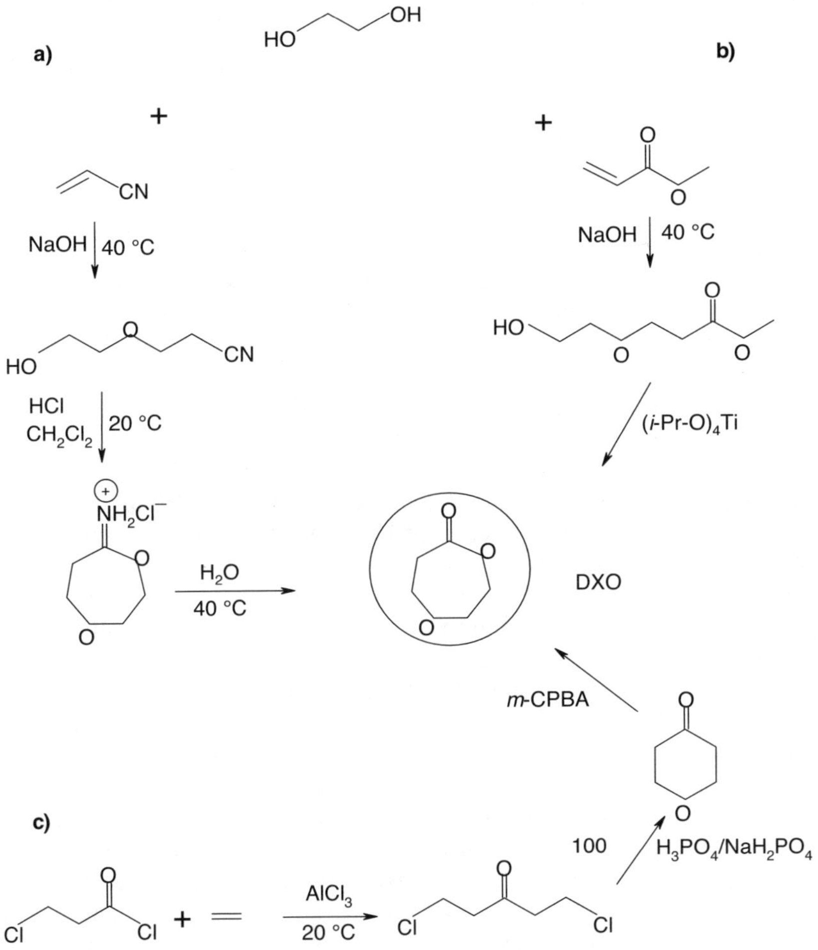

Scheme 3. Various synthetic routes for the preparation of 1,5-dioxepan-2-one

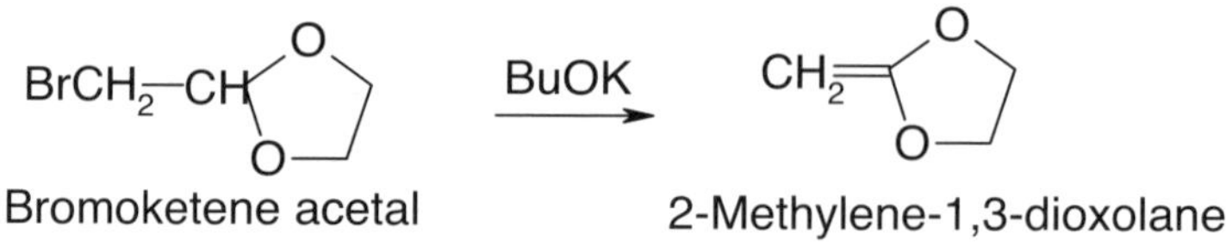

Scheme 4. Synthesis of cyclic ketene acetal

The yield of DXO based on THP was 80%. However, the synthesis of THP is the limiting step in this reaction sequence due to the low yield obtained in the cyclization reaction of 1,5-dichloropentan-3-one. The overall yield of DXO was about 48%, which is an improvement over other known synthetic routes [57].

Baeyer-Villiger oxidation of cyclohexanone is used for the preparation of ε-CL.

Cyclic ketene acetals such as 2-methylene-1,3-dioxolane are prepared by the dehydrohalogenation of the corresponding haloketene acetal (Scheme 4) [58, 59].

2.2.2
Polymerization

Carothers and colleagues were the first to explore the ROP of lactones. Many research laboratories have now been involved in this research area. The ROP of lactones is the method of choice for the production of biocompatible and biodegradable polyesters. Lactones are ambidentate and the polymerization may proceed by either alkyl-oxygen or acyl-oxygen scission. Evidence in favor of both types of scission is reported in the literature.

Thermodynamic data for small and medium sized lactones show that the entropy change during polymerization is negative. Thus, the driving force for polymerization is the negative change of enthalpy. The ROP of highly strained three- and four-membered rings is a favorable reaction – the driving force being the release of angular strain (Bayer's strain). The presence of substituents at the ring carbons further increases the strain and thus increases the exothermicity of the reaction. In medium sized rings, such as a 7-membered ring, the relief of intramolecular crowding (transannular strain) is the driving force. The ΔG_p^0 values for β-propiolactone, γ-butyrolactone, δ-valerolactone, ε-caprolactone, and α,α-dimethyl-β-propiolactone have been reported as –60.2, +12.6, –8, –12.8, and –84.02 kJ/mol, respectively. The ΔG_p^0 is positive for γ-BL, indicating that the ROP of γ-lactone is thermodynamically unfavorable because of the stability of the 5-membered ring [60, 61]. However, by use of drastic reaction conditions, polymerization may be carried out. For example, poly(γ-BL) of Mn 35,000 could be prepared under a pressure of 2×10^4 atm at 160 °C [62]. γ-BL forms high molecular weight copolymers (Mn≈10^4–10^5) with other lactones, with up to 50 mol % of repeating units derived from γ-BL.

The polymerizations are generally carried out in bulk or in solution (THF, dioxane, toluene, etc.). The dispersion polymerization of ε-CL using a mixture of 1,4-dioxane and heptane and surface-active agents yields a polymer in the form of microspheres with a narrow molecular weight distribution [63].

A few lactones polymerize spontaneously on standing or on heating. Most do so in the presence of catalysts. ROP can be performed with a large number of initiators to form high molecular weight products. Many organometallic compounds, such as oxides, carboxylates, and alkoxides, are effective initiators for the controlled synthesis of polyesters using the ROP of lactones [64, 65]. Different initiation mechanisms can be divided into free radical, anionic, carbocationic, zwitterionic, coordinative mechanisms, or mechanisms based on active hydrogen species [16]. The highest yields and molecular weights have been obtained mainly by the anionic and coordinative ROP [66].

Free radical initiators or active hydrogen compounds such as amines or alcohols are not very effective initiators for the polymerization of lactones. Polyesters of low molecular weight are produced by these techniques. For example, copolymerization of various lactones in the presence of water at 200 °C proceeded via a hydrolysis followed by the polycondensation reaction of the hydroxy acid, giving low molecular weight products [67–69]. Low molecular weight ($\approx$10,000) tri-block copolymer (CL-b-EO-b-CL) has been prepared from ε-CL and α,ω-hydroxy-terminated poly(ethylene glycol) ($Mn \cong 10^{-3}$) by carrying out the polymerization at 165 °C for several hours in the absence of catalysts [70].

Enzymatic polymerization of lactones is a promising approach and has been investigated by several workers [45, 46, 71–78]. Poly(ε-CL) with Mn=14,500 and a molecular weight distribution of 1.23 has recently been reported using *Pseudomonas* sp. lipase as the catalyst [71]. A complex mechanism involving both ring-opening and linear condensation polymerizations has been proposed for the enzymatic polymerization of lactones.

High molecular weight polyesters using ROP have been prepared by the free radical polymerization of cyclic ketene acetals. In the following text, only the methods yielding high molecular weight polyesters are described.

2.2.2.1
Free Radical Polymerization

Free radical polymerization of cyclic ketene acetals has been used for the synthesis of poly(γ-butyrolactone), which cannot be prepared by the usual lactone route due to the stability of the five-membered ring. The polymerization of 2-methylene-1,3-dioxalane at high temperatures (above 120 °C) gave a high molecular mass polyester [59, 79]. Only 50% of the rings opened when the polymerization was carried out at 60 °C, and this led to the formation of a random copolymer. The presence of methyl substituents at the 4- or 5-position facilitated the reaction. The free radical initiators generally used in such polymerizations are *tert*-butyl hydroperoxide, *tert*-butyl peroxide, or cumene hydroperoxide. The various steps involved are described in Scheme 5 [59].

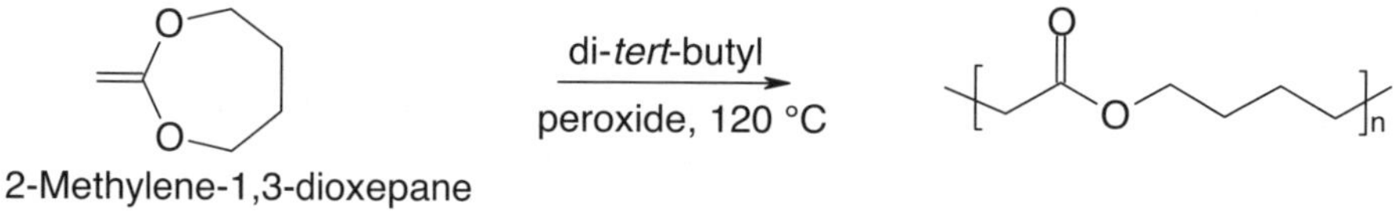

Scheme 5. Ring-opening polymerization of cyclic ketene acetal

2.2.2.2
Anionic Polymerization

The effective initiators for the anionic polymerization of lactones are alkali metals, alkali metal alkoxides (such as lithium, sodium, or potassium alkoxides), alkali metal naphthalenide complexes with crown ethers, and alkaline metals in graphite (e.g., potassium graphite), etc. Depending upon the reaction conditions, the type of initiators, and the monomers, the polymerization may proceed by a living or a non-living mechanism. Chain growth (propagation) takes place by acyl-oxygen bond scission leading to the formation of alkoxide end groups. Thus, the CH_3OK-initiated polymerization of lactones yields methyl ester and hydroxy end groups after termination with water. If alkyl-oxygen cleavage had taken place then methoxy and carboxyl end-groups would have been formed (Scheme 6).

Anionic polymerizations are sometimes associated with termination and transfer reactions. Huge amounts of cyclic oligomers, formed as a result of a backbiting reaction, were observed when ε-CL was polymerized using potassium *tert*-butoxide. However, in the presence of lithium *tert*-butoxide in an apolar solvent (benzene), oligomer formation was significantly decreased. The polymerization proceeds by the acyl-oxygen scission (Scheme 7). Smaller cyclic monomers such as β-PL do not form cyclic monomers to any great extent.

The depolymerization reaction is suppressed in the presence of a more associated alkoxide such as lithium alkoxide. γ-Lactones are difficult to polymerize by ROP. However, anionic polymerization of bicyclic bis(γ-lactones) [76] can be carried out according to Scheme 8.

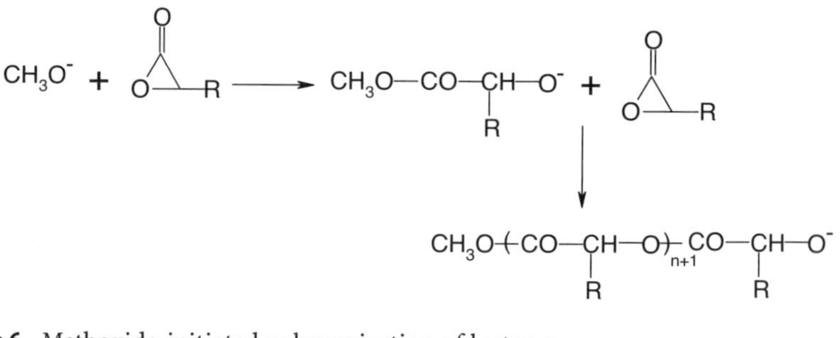

Scheme 6. Methoxide-initiated polymerization of lactones

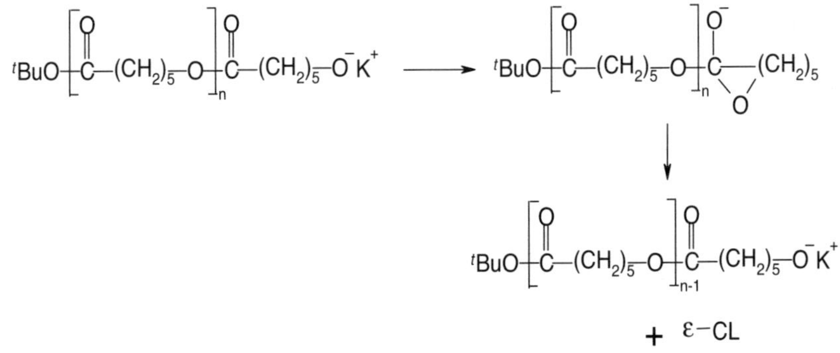

Scheme 7. Back-biting reactions during the polymerization of ε-caprolactone

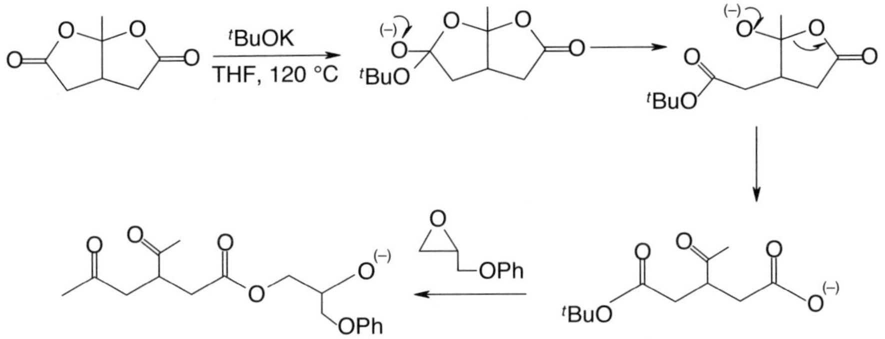

Scheme 8. Polymerization of bis(γ-lactones)

β-PL, unlike other lactones, undergoes polymerization with weakly nucleophilic initiators such as metal carboxylates, tertiary amines, phosphines, and a variety of other initiators [81–83]. This is primarily due to the high ring-strain in the four-membered ring. Pyridine and other tertiary amines initiate the anionic polymerization via a betaine that rapidly transforms into a pyridinium salt of acrylic acid. In order to minimize the chain transfer reactions, the polymerization is performed at a temperature between 0 and 10 °C (Scheme 9).

Initiation by strong nucleophiles occurs in the cases of acyl-oxygen as well as alkyl-oxygen cleavage, whereas with weak nucleophiles, ring-opening occurs at the alkyl-oxygen bond. Similarly, polymerization of pivalolactone initiated with tertiary amines or phosphines is a fast process giving high molecular weight products.

A high molecular weight β-PL is obtained by the homogeneous polymerization of β-PL in THF using metallic potassium and 18,6-crown ether. The initiation step is believed to involve the cleavage of the σ-bond between the -CH$_2$-CH$_2$- of β-PL instead of the acyl-oxygen or alkyl-oxygen bond (Scheme 10).

However, in most cases, propagation takes place through carboxylate chain-ends, although coexistence with alkoxide chain ends in the beginning of ROP

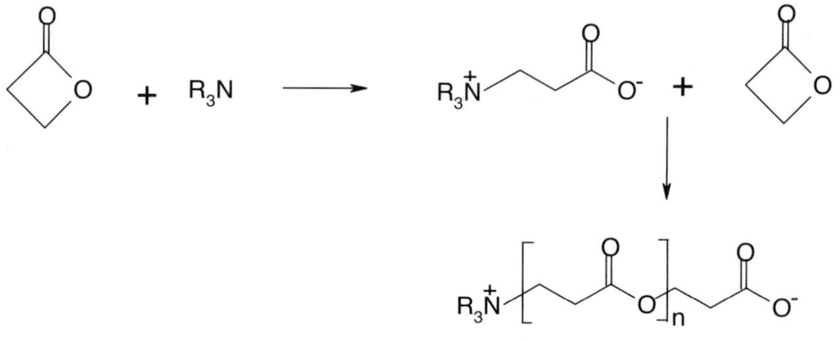

Scheme 9. Polymerization of β-propiolactone

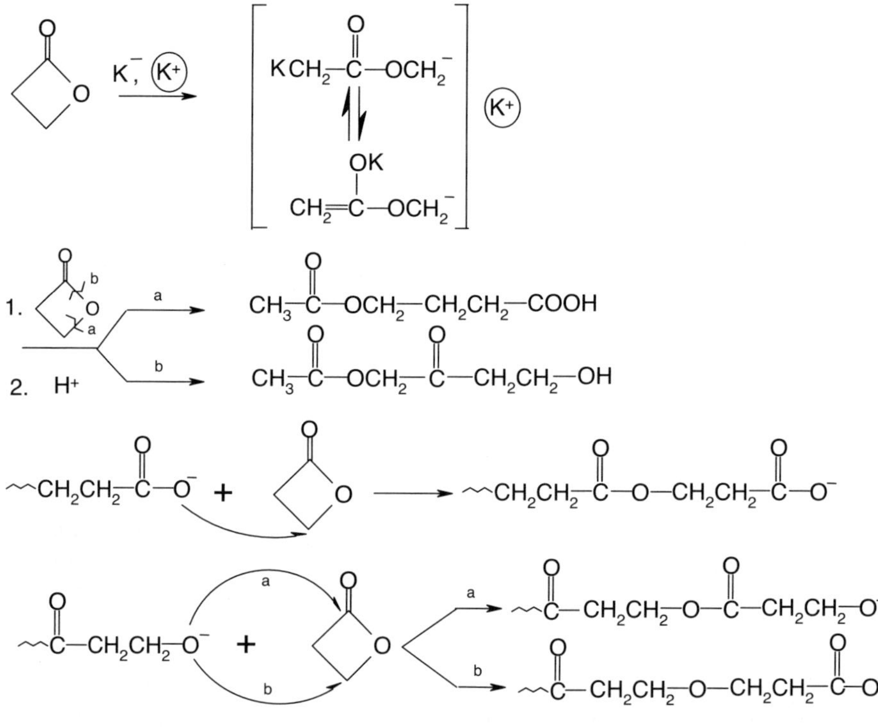

Scheme 10. Homogeneous polymerization of β-propiolactone using potassium and a crown ether

has been reported [84]. In the potassium methoxide-initiated polymerization of β-PL, terminal unsaturation was observed by Jedlinski et al. [85] and has been explained on the basis of the following mechanism (Scheme 11).

Both intra- and intermolecular transesterification reactions in anionic polymerization have been observed [82, 86–87]. Intercalation of alkaline metals in

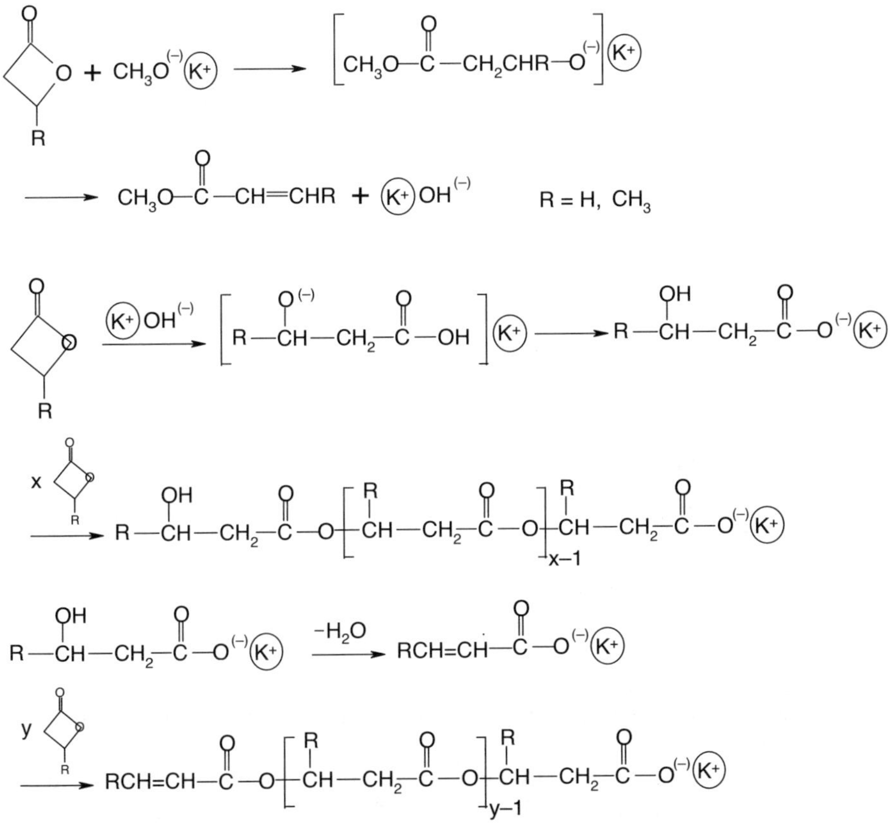

Scheme 11. Potassium methoxide-initiated polymerization of β-lactones

graphite reduces the oligomer formation in solution polymerization, which takes place within the interlamellar spacing of the graphite.

The anionic polymerization of DXO using t-BuLi or CH_3Li in THF or toluene solution was unsuccessful and only oligomers could be obtained [88].

2.2.2.3
Carbocationic Polymerization

Carbocationic polymerization is not as useful as anionic polymerization for obtaining high molecular weight polyesters, due to the occurrence of intramolecular transesterification (cyclization), proton and hydride transfer reactions. High molecular weight polymers are obtained with strained monomers such as β-propiolactone. Initiators for the ROP of lactones include protonic acids (HCl, RCOOH, RSO_3H, etc.), Lewis acids ($AlCl_3$, BF_3, $FeCl_3$, $ZnCl_2$, etc.), stabilized carbocations ($ET_3O^+BF_4^-$) and acylating agents $(CH_3CO)^{(+)(-)}OCl_4$. Okamoto [89, 90] has reported a living carbocationic ROP of ε-CL, δ-VL, and β-BL initiated by

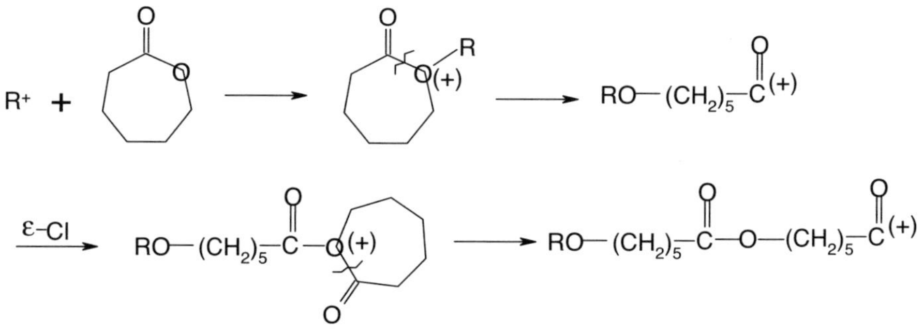

Scheme 12. Carbocationic polymerization of ε-caprolactone

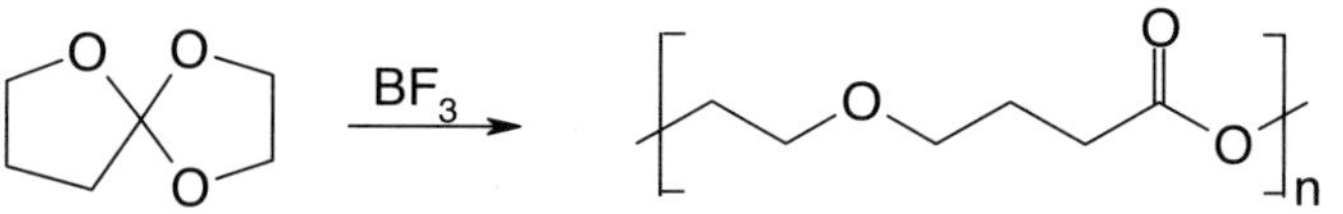

Scheme 13. Polymerization of 1,4,6-trioxaspiro[4.4]nonane

$Et_3O^+[PF_6]^-$ in bulk and in the presence of an alcohol. Proton transfer to monomer, followed by ring-opening and insertion into the chain (Scheme 12) initiates the reaction.

Cationic polymerization of DXO in bulk and in solution using Lewis acids as initiators ($SnCl_4$, $FeCl_3$, $AlCl_3$, BCl_3, and $BF_3 \cdot OEt_2$) has been reported by Albertsson and Palmgren [91]. $BF_3 \cdot OEt_2$ was found to be a good initiator for both bulk and solution polymerization. The molecular weights of PDXO increased with temperature up to 70 °C but a further increase in temperature resulted in a decrease due to back-biting reactions. A Mw of 10,900 was obtained when the reaction was carried at 70 °C for 24 h using a $[M]_o/[I]_o$ ratio of 500.

The BF_3-initiated ROP of 1,4,6-trioxaspiro[4.4]nonane yielded the poly(ether ester) (Scheme 13).

2.2.2.4
Coordinative Ring-Opening Polymerization of Lactones

Coordinative initiation differs from ionic polymerization in that the propagating species consists of a covalent bond species. This generally reduces the reactivity and the polymerization rate. Decreased reactivity also leads to fewer amounts of side reactions and the often-living ROP of lactones may take place under these conditions. Chedron, in the early 1960s, showed that some Lewis acids, such as triethylaluminum and water or ethanolate of diethylaluminum, were effective initiators for lactone polymerizations. Tin(IV) alkoxides and phenoxides, [92, 93] aluminum alkoxides, mainly aluminum *iso*-propoxide, and soluble

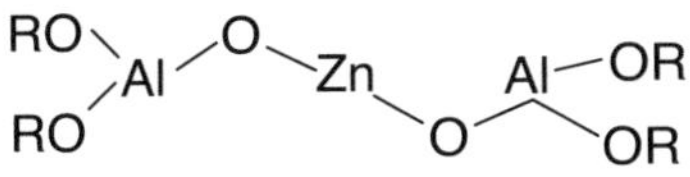

$$\text{RO}\diagdown_{\text{Al}}\diagup\text{O}\diagdown_{\text{Zn}}\diagdown_{\text{O}}\diagup^{\text{Al}}\diagdown_{\text{OR}}^{\text{OR}}$$
$$\text{RO}\diagup\phantom{\text{Al}}\phantom{\text{O}}\phantom{\text{Zn}}\phantom{\text{O}}\phantom{\text{Al}}$$

Scheme 14. Structure of bimetallic μ-oxo-alkoxides

bimetallic μ-oxo-alkoxides, essentially of zinc and aluminum, have generated a lot of interest because of their versatility as initiators [94–98] (Scheme 14).

Block copolymers of ε-CL with D,L-lactide, styrene, or butadiene have been synthesized using these initiators. Efficient and versatile initiators based on α,β,γ,δ-derivatives of tetraphenylporphinato-aluminum for the polymerization of ε-CL, β-lactones, δ-lactones, and lactides have been reported [99, 100].

Lanthanide compounds such as yttrium and lanthanum alkoxides [101, 102] as well as Sm and Lu complexes, have been reported to yield high molecular weight polyesters under mild conditions. Rapid polymerization of ε-CL at room temperature was reported when yttrium alkoxide was used as an initiator [103, 104]. Stevels et al. [105, 106] have used *in situ*-generated yttrium alkoxides as initiators for ε-CL and δ-VL. Use of bulky coordinated groups reduced the transesterification reaction, and polymers with narrow MWD could be obtained [107–109]. The performance of these initiators in terms of yield, molecular weight, polydispersity, and stereoregularity is affected by the ligands and by the oxidation state of the respective rare earth metals [110]. High selectivity with minimal side-reactions has been demonstrated in the ROP of cyclic esters with aluminum- and lanthanide-based initiators [111–113].

Aluminum-alkoxide-initiated polymerization of lactones proceeds according to a "coordination-insertion" mechanism, which involves acyl-oxygen bond cleavage of the monomer and insertion into the aluminum-oxygen bond of the initiator. The coordination of the exocyclic oxygen to the metal results in polarization and this makes the carbonyl carbon of the monomer more susceptible to nucleophilic attack. Transesterification reactions may also take place at elevated temperatures. The controlled synthesis of telechelic polymers, block and graft copolymers, and polymers of varying architecture is possible using aluminum alkoxides of different structures and functionalities [64, 65].

The following mechanism has been suggested for the polymerization of 1,5-dioxepan-2-one in bulk or in THF or toluene solution using aluminum isopropoxide as an initiator (Scheme 15).

Aluminum isopropoxide coordinates to the exocyclic carbonyl oxygen, and the acyl-oxygen cleavage yields an isopropyl ester end group. Termination of the growing chain with diluted Hcl converts the propagating end to a hydroxy group. A narrow molecular weight distribution and an increase in (DP) with increasing $[M]_o/[I]_o$ ratio indicated the living character of this polymerization. A poly(DXO) of Mn=17,500 could be obtained by polymerization in THF at 0 °C [95]. The reaction follows first order kinetics with respect to monomer and initiator, with a rate equation

$R_p = k [M] [I]$,

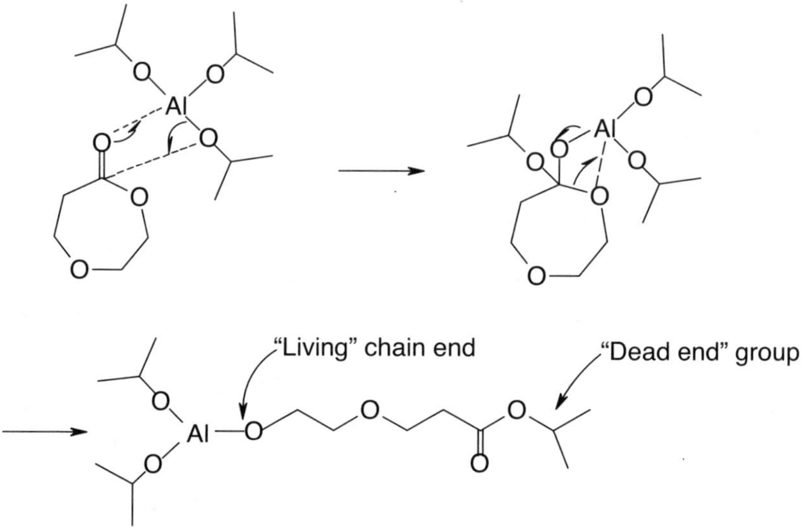

Scheme 15. Aluminum isopropoxide initiated polymerization of 1,5-dioxepan-2-one

where R_p=rate of propagation and k is the absolute rate constant equal to 5.65 L mol^{-1} min^{-1} at 0 °C in THF for DXO and 36.6 L min^{-1} mol^{-1} at 0 °C in toluene for ε-CL.

The living ROP of ε-CL is usually initiated by aluminum isopropoxide, [Al(O^{i}Pr)$_3$], in toluene at 0–25 °C. Under these conditions this initiator exists as an aggregate of trimers and tetramers. However, freshly distilled Al(O^{i}Pr)$_3$ consists mainly of trimers, and is a more reactive initiator for ROP. The initiation rate is high compared to the rate of propagation so that a narrow molecular weight distribution is obtained in the polymer. There is no termination reaction and 3 chains grow per Al atom. Block polymers have been prepared by sequential polymerization of ε-CL (monomer A) and DXO (monomer B) using Al(O^{i}Pr)$_3$ as an initiator in THF at 0 °C to yield AB or BA di-block copolymers [95].

The A-B di-block copolymer of ε-CL and oxepan-2,7-dione has been synthesized using aluminum isopropoxide as initiator [114] (Scheme 16). In order to prepare the ABA tri-block copolymer, a difunctional initiator [Et$_2$AlO(CH$_2$)$_4$OAlEt$_2$] was used to polymerize B followed by the addition of monomer A. However, the rate of polymerization was lower than in the Al(O^{i}Pr)$_3$-initiated system. Increasing the temperature to 70 °C increased the rate but a broadening of MWD was observed due to intramolecular back-biting reactions and intermolecular transesterification reactions. The addition of 1 equiv. of pyridine with respect to Al increased the polymerization rate and reduced the MWD from 1.95 to 1.25 [95].

In β-lactones, scission of either the acyl-oxygen bond or the alkyl-oxygen bond may take place leading to the formation of alkoxide- or carboxylate-growing chains [115]. Methylene chloride end groups were observed in the ZnCl$_2$-in-

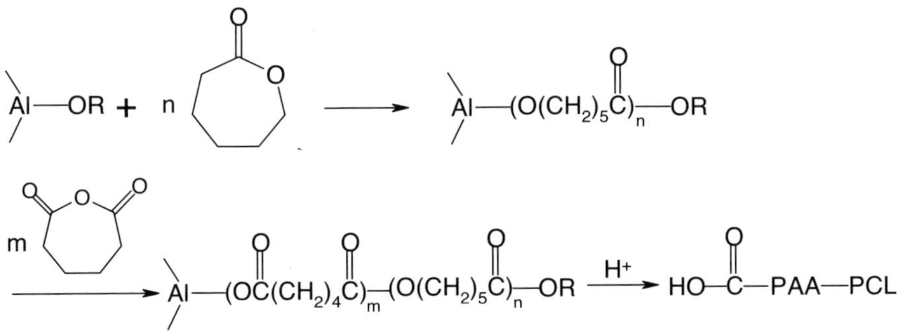

Scheme 16. Reaction of ε-caprolactone and oxepan-2,7-dione

itiated polymerization of ε-CL in xylene, and this supports the coordination-insertion mechanism [116].

Carboxylates are less nucleophilic than alkoxides and are considered to behave more like a catalyst rather than an initiator. Metal carboxylates such as tin(II) 2-ethylhexanoate, $Sn(Oct)_2$, are used together with active hydrogen compounds (e.g., alcohols) as co-initiators [57, 117]. If no active hydrogen compound is added, the actual initiating species may be hydroxy-containing impurities [16, 118]. $Sn(Oct)_2$ is a very effective and versatile catalyst which is commercially available, is easy to handle, and is soluble in common organic solvents and lactones. The mechanism of ROP in the presence of tin(II) 2-ethylhexanoate [$Sn(Oct)_2$] has recently been examined by Kowalski et al., [119–121] and by Kricheldorf et al. [122]. The active species of the ε-CL polymerization in THF was identified as $OctSn[O(CH_2)_5C(O)]_nOR$ which suggests that polymerization proceeds on a tin(II) alkoxide bond formed from $Sn(Oct)_2$.

Recent studies of the tetrabutyltin-initiated polymerization of ε-CL have indicated that Bu_2SnO dissolved in Bu_4Sn is the main initiator [123]. Almost 100% conversions are observed in the living macrocyclic polymerization of ε-CL in bulk at 80 °C with 2,2-dibutyl-2-stanna-1,3-dioxepane as initiator [124].

Albertsson et al. [55, 56, 95, 114, 125–138] have done extensive work on the homo- and copolymerizations of lactones in bulk as well as in solutions using ROP. In bulk polymerization temperatures in the range of 100–150 °C were used while in solution polymerization, the temperature was kept low, 0 to 25 °C, to minimize side-reactions such as intra- and intermolecular transesterification reactions. Only oligomers were formed when DXO was (co)polymerized using an ionic initiator. Poly(DXO) of high molecular weight (>150,000) was obtained using tin(II) 2-ethylhexanoate [127].

α-Methacryloyl-ω-hydroxy-poly(ε-CL) macromonomer has been synthesized using $Sn(Oct)_2$ as catalyst and 2-hydroxyethyl methacrylate as initiator. This macromonomer had a methacryloyl group at one end and a hydroxy group at the other. The molecular weight and polydispersity of the macromonomer depended on the concentrations of $Sn(Oct)_2$ and HEMA as well as the molar ratio

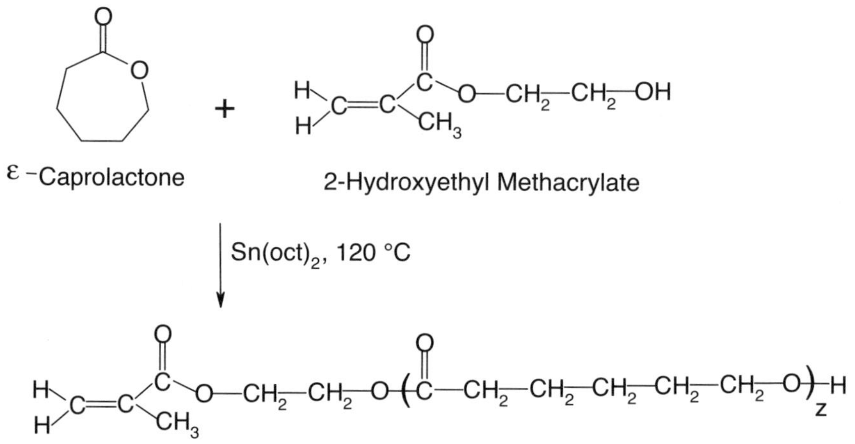

Scheme 17. α-Methacryloyl-ω-hydroxyl-poly(ε-caprolactone) macromonomer

of Sn/OH [139] (Scheme 17). Copolymerization of this macromonomer with HEMA using AIBN yielded a graft copolymer having a poly(HEMA) main chain and poly(ε-CL) as graft chain.

High molecular weight poly(DXO) has been prepared at 110 °C using tin(II) 2-ethylhexanoate as catalyst [127]. Transesterification and degradation occurred above 130 °C.

The reactivity ratios in the copolymerization of DXO and L-LA were $r_{(DXO)}$= 0.1 and $r_{(L-LA)}$=10 [133] with Sn(Oct)$_2$ as catalyst. Tin catalysts have been shown to cause transesterification, especially at elevated temperatures (>140 °C) [140]. This could cause a reshuffling of the initially formed sequences to a more random structure. Lactic acid units are more sensitive to transesterification than DXO.

Reactivity ratios in the random copolymerization of DXO and ε-CL have been reported as r_{DXO}=1.6 and $r_{ε-CL}$=0.6 [132]. The copolymer of DXO and L-LA is expected to have a more block-like nature whereas poly(DXO-co-ε-CL) is expected to be completely random. Reactivity ratios for the copolymerization of DXO and δ-VL were determined at 110 °C as r_{VL}=0.5 and r_{DXO}=2.3 [138]. Thus, DXO is consumed faster in copolymerization with ε-CL and δ-VL. At high conversion, depolymerization of poly(δ-VL) occurred, resulting in a lower molecular weight and variations in copolymer composition. Copolymerization of 1,3-dioxan-2-one (TMC) and ε-CL could be carried out using SnOct$_2$, Bu$_2$SnO, but Bu$_3$SnCl was not effective. The reactivity ratios in this copolymerization were r_{TMC}=0.17 and $r_{ε-CL}$=2.47. The copolymer formed is richer in ε-CL and has a block-like character [128].

In the random copolymerization of δ-VL and ε-CL using [SmMe(C$_5$Me$_5$)$_2$ (THF)], δ-VL is found to be more active [$r_{(VL)}$=2.82 and $r_{(ε-CL)}$=0.20]. The highly strained β-PL has a higher reactivity than ε-CL and δ-VL in copolymerizations [$r_{(β-PL)}$=16.6, $r_{(ε-CL)}$=0.22, and $r_{(β-PL)}$=6.73, $r_{(δ-VL)}$=0.32] [110].

The Sn(Oct)$_2$ catalyst is generally active at elevated temperatures, leading to some intermolecular and intramolecular transesterification reaction [140]. Recently, a new catalyst system for the ROP of lactones has been reported, based on tin(II) or scandium(III) trifluoromethanesulfonate [141, 142]. These catalysts are very versatile, and highly selective, and they can be used under mild conditions.

Homopolymers and random copolymers of ε-CL (r_1=0.37) and D,L-lactide (r_2=10.8) have been reported using lanthanide halides as initiators [107]. The addition of epoxides or N-trimethylsilyltrialkyl(or -aryl)phosphinimines enhances the reactivity of the lanthanide halides [143]. High molecular weight poly(ε-CL) was obtained at room temperature using an SmBr$_2$/Sm system [143, 144]. Poly(p-xylylene-g-ε-CL) graft copolymers have been prepared by *in situ* reductive coupling of aldehydes using SmI$_2$ [145, 146]. Block copolymers have been prepared using halogen-bridged Sm(III) complexes.

Aluminum isopropoxide has been used for the preparation of block copolyesters [147, 148]. Tri-block poly(ε-CL-b-DXO-ε-CL) was prepared by the sequential addition of different monomers to a living polymerization system initiated with aluminum isopropoxide in THF or toluene solution [95]. An alternative route for the preparation of the tri-block copolymer was to react the diblock poly(ε-CL-b-DXO) containing an –OH functionality at the chain end using a difunctional coupling agent such as isocyanate or acid chloride (Scheme 18). However, the molecular weights were low and full conversion of monomers was not achieved.

The addition of a Lewis base (pyridine) did not significantly alter the polymerization rate in THF, although a small increase in rate was observed in toluene,

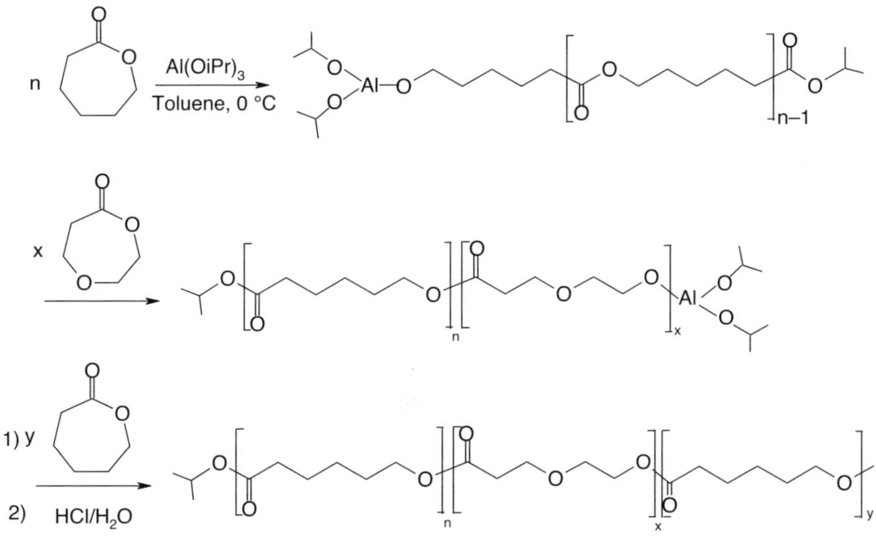

Scheme 18. Preparation of tri-block poly(ε-caprolactone-b-1,5-dioxepan-2-one-ε-caprolactone)

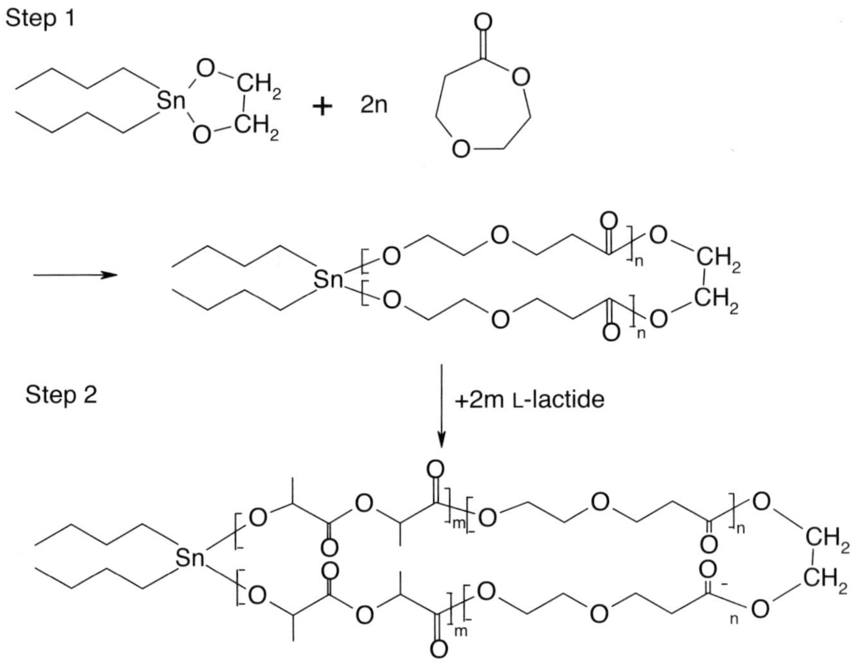

Scheme 19. Preparation of poly(L-lactide-b-1,5-dioxepan-2-one-b-L-lactide)

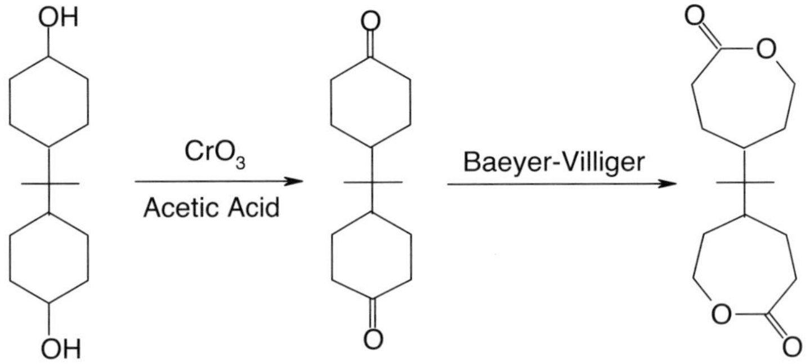

Scheme 20. Preparation of a cross-linker for 1,5-dioxepan-2-one

especially at the later stages of polymerization. The optimal conditions were found to be polymerization in toluene solution at 25 °C with the addition of 1 molar equivalent of pyridine (with respect to initiator). The poly(ε-CL) block should first be synthesized at 0 °C and the following two blocks at 25 °C, in order to increase the polymerization rate.

The preparation of poly(L-lactide-b-DXO-b-L-lactide) has recently been reported using controlled ROP in a two-stage process: polymerization of DXO using a difunctional initiator: 1,1,6,6-tetra-n-butyl-1,6-distanna-2,5,7,10-tetraoxacyclodecane, and the addition and polymerization of L-lactide in chloroform at 60 °C [148] (Scheme 19).

Poly(ε-CL-g-acrylamide) copolymers could be readily prepared using electron beam irradiation and Fe^{2+} ions provided by Mohr's salt as efficient terminators for grafted polyacrylamide chain radicals. Varying the Fe^{2+} ion concentration [149] enabled the average molecular weight of grafted chains to be varied within wide limits. Poly(DXO) and chemically cross-linked poly(DXO) (Scheme 20) could be grafted with acrylamide under similar conditions [150, 151].

2.3
Functional Polyesters

It is obvious from the above discussion that ROP makes it possible to prepare polyesters with defined molecular weights and functionalized end groups as well as random, block, and graft copolymers. The properties of polyesters (such as hydrophilicity, rate of biodegradation, etc.) can also be tailored by introducing functional groups along the polymer chains. Very few such functional polyesters have been reported in the literature [152, 153]. Homopolymerization of 2-allyl-ε-caprolactone or 6-hydroxy-8-enoic acid lactone in bulk at 110 °C using $Sn(Oct)_2$ as catalyst and monofunctional (benzyl alcohol), bifunctional, or multifunctional co-initiators such as the benzyl ester of bis-MPA (bifunctional) or the hexahydroxy functional dendrimer of the bis-MPA (hexafunctional) yielded polyesters with pendant allyl groups which could be easily epoxidized, brominated, or silylated. Homopolyesters, as well as random and block copolymers having pendant functional groups have been reported [154, 155] (Scheme 21).

2.4
Hyper-Branched Polyesters

Hyper-branched aliphatic polyesters are different from linear polyesters because of their unique mechanical and rheological properties, which can easily be tailored by changing the nature of the end group [156–159]. Both dendrimers and hyper-branched macromolecules are prepared from AB_2 monomers, using dif-

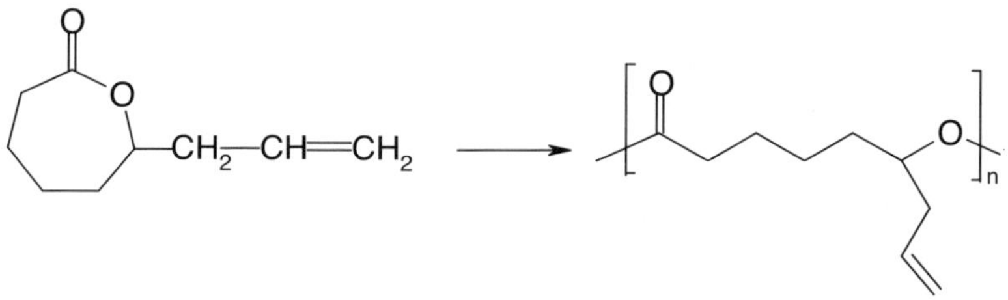

Scheme 21. Preparation of poly(6-allyl-ε-caprolactone)

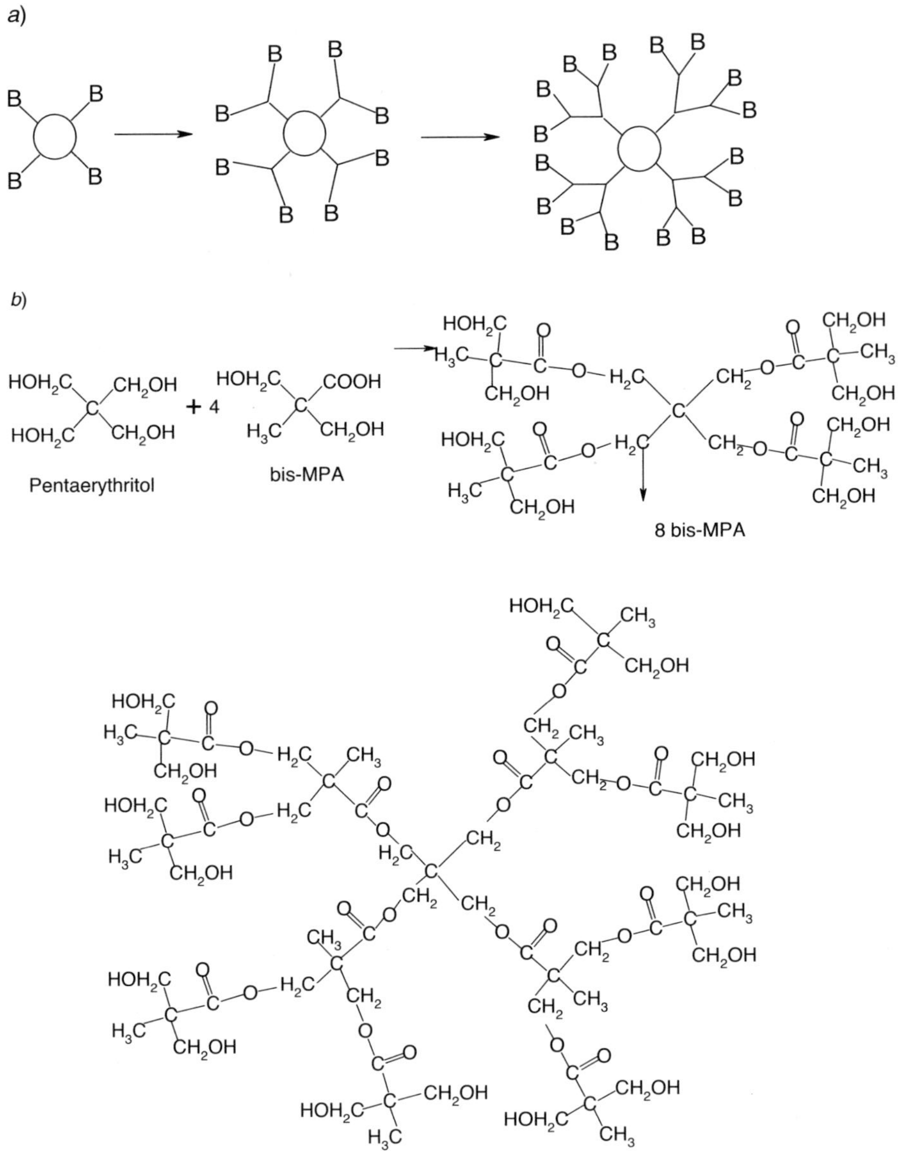

Scheme 22. *a)* The divergent growth approach for the preparation of dendritic polyesters. *b)* The divergent approach for the preparation of dendrimer having 16-hydroxy groups

ferent synthetic strategies. Dendrimers are produced by a stepwise series of reactions using 2 approaches:
(a) the divergent growth approach, which starts from a multifunctional core molecule and proceeds radially outward (Schemes 22 and 23),

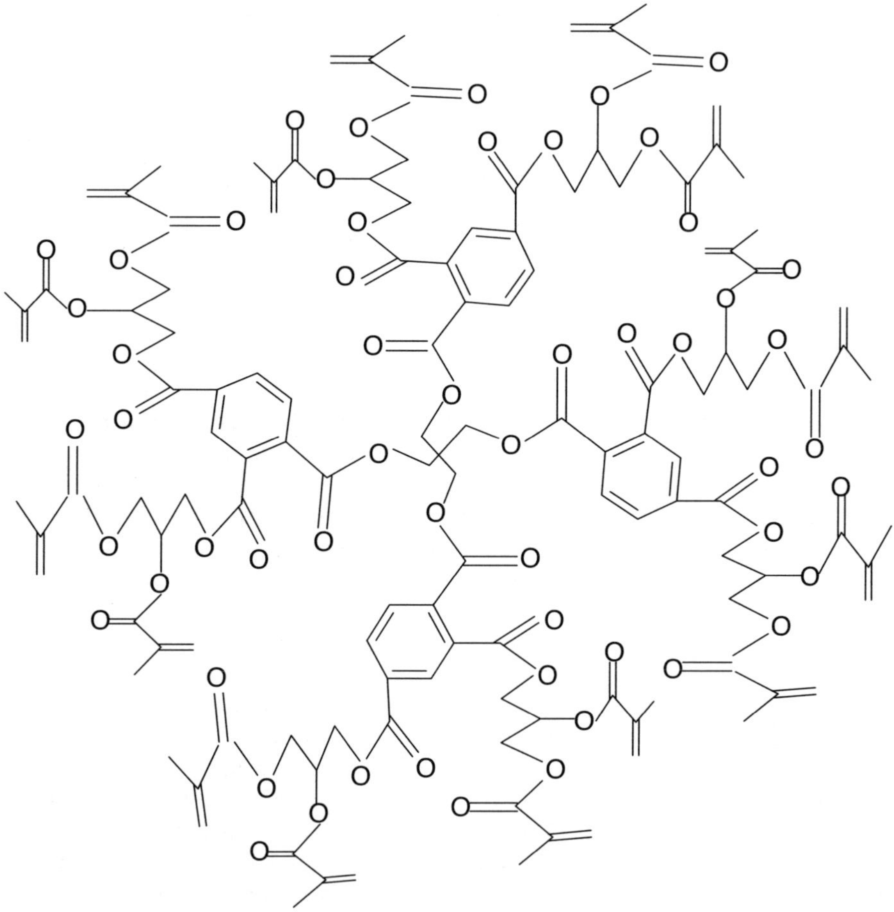

Scheme 23. Polyesters having a pendant double bond prepared by the divergent growth method (adapted from [156])

(b) the convergent growth approach, in which the synthesis starts at the periphery and well-defined dendrons are prepared, followed by final coupling to a multifunctional core [160–165].

Hyper-branched polymers are prepared in a single-step polymerization from AB_x monomers. Thus, a perfectly branched structure is present in dendrimers, whereas irregular branching is present in hyper-branched polymers. Aluminum alkoxide-based initiators or tin-based catalysts have been successfully used for the preparation of, hyper-branched [160–162, 166–168], dendrimer-like star polymers [160], and star-shaped polymers. The first and second generations of the benzyl ester of 2,2-bis(hydroxymethyl)propionic acid (bis-MPA) are effective initiators for the ROP of lactones (ε-CL) in the presence of $Sn(Oct)_2$. The requisite AB_2 and AB_4 macromonomers and hyper-branched polyesters can be prepared by deprotection of the benzyl ester group from the initiator.

A commercially available hyper-branched polyester derived from bis-MPA was used as the multifunctional initiating core for the ROP of ε-CL and this led to the synthesis of hybrid dendritic linear star polymers. The reactivities of the chain-end hydroxymethyl groups in the dendrimer were significantly greater than in the isomeric hyper-branched case.

Lipase catalysis has been used for the synthesis of a poly(ε-CL)-monosubstituted first generation dendrimer [169].

3
Physical Properties

The physical properties of aliphatic polyesters depend on several factors such as the composition of repeat units, flexibility of the chain, presence of polar groups, molecular mass, degree of branching, crystallinity, orientation, etc. Short-chain branches reduce the degree of crystallinity of polymers while long chain branches lower the melt viscosity and impart elongational viscosity with tension-stiffening behavior. The properties of these materials can further be tailored by blending [170], and copolymerization [133, 144] or by a change in the macromolecular architecture (i.e., hyper-branched polymers, star-shaped or dendrimers, etc.) [171].

BIONELLE is a white crystalline thermoplastic with a melting ranging from 90–120 °C, Tg ranging from –45 to –10 °C, and a density of about 1.25 g/cm^3. It has excellent processability and can be processed on conventional equipment at a temperature of 160–200 °C into various molded products such as injected, extruded, and blown materials [11]. The carbon content in the glycol and acid affects the melting behavior significantly. The melting points of poly(ethylene adipate), PEA, and poly(butylene succinate), PBS, have been reported as 47.2 and 115.9 °C, respectively [172]. The structural change in these polymers by copolymerization or by incorporation of ethyl or octyl branches affects their properties (Table 2).

Table 2. Effect of structure on the properties of aliphatic polyesters

Sample	Tg (°C)	Tm (°C)	σ (Mpa)	E (Mpa)	ε (%)
PBS	–36.6	115.9	37.2	1901	47.0
PEA	–46.4	47.2	13.2	312.8	362.1
PEAB04	–47.0	45.6	12.1	254.4	248.3
PEAB15	–48.1	38.4	5.3	119	71
PEAD05	–46.5	46.1	5.2	167.5	245.3
PEAD08	–46.7	44.1	4.3	172.3	114.4
PEAD12	–47.1	38.6	3.5	174.9	4.5

PEA=poly(ethylene adipate); PBS=poly(butylene succinate), D=octyl branches and B=ethyl branches; the numerals indicate mol % of branches in the copolymer Data adapted from [172]

Poly(L-LA) is a semicrystalline polymer with a Tg of 61 °C and Tm of 174 °C. The crystal structure of PLLA is pseudo-orthorhombic [173]. It is a white fibrous material.

Poly(β-PL) has a Tg of –15 °C and a Tm of 83 °C [174]. The degree of crystallinity depends on the method of preparation and ranges from 40–60%. Proper drying of the sample leads to an increase in Tg to –4 °C [175]. A tensile strength of 103 MPa and a tensile modulus of 1.59 GPa have been reported for unoriented strips.

Poly(ε-CL) is a semicrystalline polymer with a degree of crystallinity of 50%, a low Tg of ≈–60 °C and a melting point of +60 °C. The high molecular weight polymer has been reported to have a Tm of 78.9 °C [176]. Injection-molded samples exhibit a modulus of 400 MPa and yield stress of 15 MPa [177]. The material can be processed without any significant molecular weight reduction by injection molding, film blowing, and extrusion. However, its rate of crystallization is lower than that of the conventional polymers. Poly(δ-VL) has physical properties similar to those of poly(ε-CL) and has a Tm of 59 °C and Tg of –67 °C, but its crystal structure differs slightly [174].

Aliphatic poly(ether-ester)s are more flexible because of the presence of ether linkages. Poly(1,4-dioxan-2-one) is a crystalline polymer with a tensile strength and elasticity similar to those of human tissue. Poly(ε-CL) and poly(DXO) resemble each other in their chemical architecture, but poly(ε-CL) is semicrystalline while poly(DXO) is an amorphous polymer with a Tg of about –37 °C.

Physical properties such as crystallinity and melting temperatures can be significantly affected by copolymerization. Copolymers of glycolic acid and lactic acid have a Tg higher than body temperature (except for samples of low molecular weight). This fact and the usually crystalline nature of these copolymers limit their applications as biomedical implant materials. By introducing DXO, biodegradable ether-ester units in the backbone of aliphatic polyesters such as poly(ε-CL) or polylactides, flexible and more pliable materials can be prepared. Copolymerization of L-lactide with DXO yielded a semicrystalline yet more flexible and ductile material than pure PLA [137, 178]. The Tg of poly(DXO-co-L-LA), ranges from –36 up to +58 °C depending upon the molar composition. Thus, the Tg and Tm of poly(L-LA-co-DXO) having 90:10 and 70:30 ratios of L-LA:DXO have been reported as 43.6, 22.1 and 161.5, 143.9 °C, respectively [179]. The block copolymer exhibited crystallinity as long as the amount of DXO did not exceed 50 weight %. The copolymers of DXO and δ-VL showed negligible crystallinity when the DXO ratio in the monomer feed was larger than 25%, while all poly(DXO-co-L-LA)s exhibited crystallinity to some extent. The Tg and Tm of these copolymers are given in Table 3.

The Tg of copolymers of ε-CL and DXO was in the range from –64 to –39 °C [131]. Both crystallinity and Tm decrease with increasing amount of DXO and about 40% of DXO comonomer units of poly(ε-CL-co-DXO) are incorporated into the poly(ε-CL) crystals. Some inclusion of DXO in the crystalline lattice of poly(δ-VL) was also indicated on the basis of crystallinity data [138].

Block copolymerization is yet another way of preparing novel, highly flexible and degradable polymers. Block and graft copolymers possess unique proper-

Table 3. Properties of poly(CL-co-DXO), poly(VL-co-DXO), and poly(LLA-co-DXO)

Sample	% DXO in copolymer	T_g (DSC) [°C]	T_m (DSC) [°C]
CD50	50	–56.8	27.8
CD60	41	–57.8	27.2
CD70	29	–55.5	36.0
CD80	18	–61.0	42.8
CD90	8	–65.6	50.5
CD100	0	–65.9	57.6
VD70	33	–56.7	28.0
VD80	25	–56.1	37.7
VD90	7	–59.9	46.0
VD100	0	–63.4	57.5
LD70	28	23.1	154.1
LD85	13	41.1	170.8
LD100	0	58.5	183.8

Data adopted from [134]

Table 4. The mechanical properties of poly(ε-caprolactone)/poly(1,5-dioxepan-2-one)/poly(ε-caprolactone) tri-block copolymers (standard deviation in parentheses)

Block length A/B/A g/mol	15 K/50 K/15 K	10 K/70 K/10 K
Tensile modulus (MPa)	31 (8.8)	21 (4.2)
Stress at yield	3.0 (0.4)	2.1 (0.1)
ε at yield %	15 (4.3)	18 (0.4)
Strength at break (MPa)	52 (6.7)	53 (30)
ε at break %	1070 (67)	1210 (67)

ties in that they combine the inherent nature of both the homopolymers. Such copolymers may be tailored to provide a range of properties from brittle plastics to elastomers. Well-defined block copolymers have been synthesized. Poly(ε-CL-b-DXO-b-ε-CL) with high molecular weight and narrow molecular weight distribution showed a melting peak at 51 °C and a Tg of –38 °C corresponding to poly(ε-CL) crystalline and poly(DXO) amorphous phases, respectively. The mechanical properties of these copolymers depend on the chemical composition (Table 4).

Blending of polymers is an attractive method of producing new materials with better properties. Blends of aliphatic polyesters, especially of poly(ε-CL), have been investigated extensively and have been the subject of a recent review paper [170]. Poly(ε-CL) has been reported to be miscible with several polymers such as PVC, chlorinated polyethylene, SAN, bisphenol A polycarbonate, random copolymers of Vdc and VC, Vdc and AN, and Vdc/VAc, etc. A single composition-dependent Tg was obtained in the blends of each of these polymers with poly(ε-CL). This is of interest as a polymeric plasticizer in these polymers. Blends of PVC and poly(ε-CL) with less than 50 wt % of poly(ε-CL) were homogeneous and exhibited a single Tg. These blends were soft and pliable because the inherent crystallinity of poly(ε-CL) was destroyed and PVC was plasticized by amorphous poly(ε-CL). The Tg of PVC (83 °C) decreased with increasing poly(ε-CL) content and was 29, 9 and –30 °C for samples containing 75:25, 50:50, and 25:75 PVC: poly(ε-CL), respectively [180]. The elasticity modulus decreased with increasing poly(ε-CL) content in amorphous systems but at high poly(ε-CL) content it increased due to the reinforcing effect of the crystalline phase. SAN copolymers containing 8–28% of AN were compatible with poly(ε-CL) in all proportions [181].

Poly(glycolic acid) and poly(3-HB) form immiscible blends with poly(ε-CL). Poly(3-HB) is an expensive polymer, which crystallizes slowly from the melts and embrittles on ageing. Poly(ε-CL), on the other hand, has a low mp and limited use. An optimal combination of cost and properties may be obtained by blending [114].

Poly(DXO) formed miscible blends with PDLLA and partially miscible blends with PLLA. This is due to a difference in the morphology of PLLA and PDLLA, the former being semicrystalline while the latter is amorphous. The Tg of homogeneous blends ranged from +48.1 (pure PDLLA) to –28.1 °C (copolymer having 29 weight % of PDLLA) [182–183]. Blend composition was a vital factor in determining the blend properties, morphology, and *in vitro* degradation.

Cross-linked poly(DXO) was obtained by ROP of DXO in the presence of a tetrafunctional bis(ε-CL) such as 2,2-bis(ε-CL-4-yl)propane (BCP) or bis(ε-CL-4-yl) (BCY). The Tg of cross-linked polymers increased with an increase in mole % of BCP or BCY [137].

The electron beam irradiation of poly(ε-CL) and cross-linked PDXO in argon yielded a secondary alkyl ether radical and a tertiary alkyl radical, respectively. When the irradiation was carried out in an air atmosphere, peroxy radicals were detected in poly(ε-CL) but not in PDXO (150). Oxygen permeabilities of poly(ε-CL), and its tri-block copolymer poly(ε-CL-b-PEG-b-ε-CL) were in the range of 10^{-10} to 10^{-9} cm^3 (STP) cm/cm^2 sec cm Hg [184].

4
Degradation

The biodegradability of a polymer very much depends on its backbone structure [185]. The most important requirement is the presence of hydrolyzable and/or oxidizable linkages in the backbone. The rates of biodegradation of polyesters are largely dependent upon the type of repetitive unit, composition, sequence length, molecular geometry, molecular weight, morphology (e.g., crystallinity, size of spherulites, orientation), hydrophilicity, surface area, and additives [15]. The homogeneous hydrolytic degradation of these polymers is generally divided into two extreme cases, the bulk erosion and the surface erosion systems. The term degradation implies bond cleavage while erosion implies depletion of material. In the case of bulk erosion, the material is lost from the entire polymer volume, whereas in surface erosion, the material is lost only from the outermost parts of the sample. Hydrophilicity and crystallinity are important factors affecting both cases. Surface erosion is proportional to surface area and a large surface area gives faster erosion. The erosion of the materials is in most cases subsequent to degradation.

The degradation can be monitored by measuring molecular weight changes, which arise due to bond cleavage, or by measuring weight loss, which is due to depletion of low molecular weight material. The hydrolytic degradation results in morphology changes, topological changes (SEM), the formation of degradation products (GC-MS), and changes in mechanical properties. Specialized chromatography (headspace GC-MS, LC, etc.) has been developed for the detection of low molecular weight compounds in degradable polymers. Degradation products using these techniques have been identified in PHA, PLA, poly(LA-co-glycolide) by Karlsson and Albertsson [186].

4.1
Homopolyesters

Takiyama and Fujimaki [187] have indicated that, for aliphatic polyesters, Mn has to be raised above 30,000 in order to obtain high mechanical strength, but biodegradability decreases above 50,000. Tokiwa et al. [188] have shown that the development of crystallinity depressed the microbial degradation and that the size of spherulites was a more important factor than crystallinity. Oriented and unoriented fibers show different degradation properties, especially with regard to changes in mechanical properties. The rate of enzymatic degradation depends on the draw ratio of fibers, suggesting that crystallinity and orientation both affect biodegradability [189].

It is believed that chain scission occurs through simple hydrolysis, but the kinetics of this hydrolysis are influenced by anions, cations, and enzymes [190]. The process is autocatalytic and the products of hydrolysis such as carboxylic groups participate in the transition state. Water preferentially enters the amorphous parts but crystalline domains are also affected [125]. The degradation of aliphatic polyesters is believed to be dominated by a hydrolytic mechanism but it is also promoted by enzymatic activities [4, 7, 191–193].

In the degradation studies of ^{14}C-labeled poly(tetramethylene adipate), a decrease in molecular weight and first an increase and later a decrease in crystallinity (DSC) were observed. These studies indicated that degradation was an abiotic hydrolysis that initially takes place in amorphous regions. The microorganisms were able to degrade the oligomers formed during the hydrolysis [42]. Oriented fibers of this polymer retained more than 50% of their original tensile strength during 100 days of abiotic hydrolysis. Drastic changes in properties of block copolymers of poly(ethylene succinate) and poly(tetramethylene glycol) having 59 mol % of polyether were observed on hydrolytic degradation at 37 °C [194]. Fibers extruded from a thermoplastic elastic block copolymer, poly(tetramethylene glycol)/poly(ethylene succinate), degraded *in vitro* at 37 °C. The hydrolytic degradation reaction, initiated in the amorphous regions, resulted in a significant decrease in tensile strength [194]. A degradation mechanism involving a combined effect of surface erosion and hydrolytic attack on the ester linkages connecting the amorphous polyether and the crystalline polyester blocks was proposed. Films of poly(ethylene succinate) eroded completely in fresh water within 10 days but then were hardly eroded in sea water after 28 days [195].

BIONOLLE is biodegradable in compost, in moist soil, in fresh water with activated sludge, and in sea water [11]. Electron beam-irradiated blends of BIONELLE and rubber showed no enzymatic degradation with lipase AK in the presence of surfactant ($MgCl_2$) and phosphate buffer (pH 7.4) at different temperatures (25–80 °C) whereas unirradiated samples containing 5% rubber degraded by up to 97% at 70 °C. The enzymatic attack occurred in both crystalline and amorphous regions of the molecule [196]. The biodegradation rates of cross-linked polyesters prepared by condensation polymerization of L- or D-malic acid and glycols having 2–6, 8–10, and 12 methylene groups using *Rhizopus delemar* lipase or activated sludge depended on the methylene content of the glycol and on the stereochemistry. The L-isomer degraded faster than the D-isomer [197]. Mochizuki and Hirami [189] observed that enzymatic degradation of films made from poly(butylene succinate-ethylene succinate) depended on the degree of crystallinity rather than on the primary structure. Studies on blown films of such copolymers revealed that the degree of orientation played a dominant role in determining the rate of surface erosion [198].

Poly(β-PL) has been hydrolytically degraded in a buffered salt solution (pH 7.2) at 37 °C [125, 175]. Oriented fibers and unoriented fibers showed different mechanical properties on degradation. The changes in tensile strength were slower for the oriented material. The molecular weight decreases rapidly during the first 50 days while the crystalline phase increases, probably due to an annealing-like effect caused by the temperature at which the degradation was performed and rapid water absorption [175].

The biodegradability of poly(ε-CL) and biocompatibility of its degradation products are well documented in the literature [199]. It is degradable in several biotic environments, such as river and lake waters, sewage sludge, farm soil, paddy soil, creek sediment, roadside sediment, pond sediment, and compost [200, 201]. Abiotic degradation of poly(ε-CL) has also been studied at different pH values and different temperatures [202, 203]. It was shown in the 1970s that poly(ε-CL) is easily biodegraded and that it is utilized as carbon source by various microbial species [204–206]. The degradation mechanism proposed is hydrolysis of the polymer to 6-hydroxycaproic acid, an intermediate of ω-oxidation, and then β-oxidation to acetyl-SCoA, which can then undergo further degradation in a citric acid cycle. On exposure to the attack of *Aurebasidium (Pullularia) pullulans*, low molecular weight poly(ε-CL) lost weight more quickly than samples with high molecular weight [204]. When polymers with a broad molecular weight distribution were degraded, the low molecular weight fraction was preferentially utilized. The degradation of poly(ε-CL) and other polyesters was investigated using *Penicillium* fungal species [205, 206]. This mold completely degraded even a high molecular weight poly(ε-CL). The molar mass decreases readily during biodegradation and this is accompanied by a broadening of the molar mass distribution [207]. Albertsson et al. observed the appearance of parallel grooves on the polymer surface during biotic degradation. No surface changes were observed in the samples exposed to an abiotic environment [208]. Biotic hydrolysis in simulated compost gave rise to holes in the poly(ε-CL) films which was the opposite of the case in traditional garden waste [209, 210].

The degradation rate of poly(ε-CL) is not affected by its shape (i.e., as a film or as microparticles) and is enhanced by the presence of enzyme lipase [211]. The degree of crystallinity of poly(ε-CL) increased with degradation indicating that preferential degradation takes place in amorphous regions [200, 211]. Recycling and the addition of processing additives reduced the degradation rates. The degradation in biotic environments, composts, and anaerobic sewage sludge was higher than in abiotic environments due to synergism between high temperature and a richer fauna of microorganisms. The enzymatic degradation rates of radiation-cross-linked poly(ε-CL) using lipase AK enzyme in a phosphate buffer solution at pH 7 were slower than that of uncross-linked poly(ε-CL) due to their network structure [212].

Microorganisms that degrade poly(ε-CL) are widely distributed in nature [213]. Fungi of the genus *Fusarium* secrete enzymes (depolymerases) which hydrolyse the insoluble polyester to water-soluble products that are used for carbon and energy. The enzyme that hydrolyses naturally occurring hydrophobic polyesters such as cutin and lipids may also attack poly(ε-CL) [214, 215]. In most environments poly(ε-CL) has been found to biodegrade more slowly than other biodegradable polymers such as poly(hydroxybutyrate-co-valerate), regenerated cellulose, and starch, etc. [216]. Its half-life is about one year *in vivo* [217].

4.2
Copolyesters

As mentioned earlier, the rate of degradation of a polyester depends on several factors, the most important being its chemical components and degree of crystallinity. Both these properties can easily be modified by copolymerization with monomers that yield polymers with large differences in Tg and/or crystallinity. One such system is the DXO/lactic acid copolymer. Lactic acid yields a homopolymer with a Tg of about 55 °C and has an amorphous or crystalline structure depending upon the stereoisomer used. Poly(DXO) is an amorphous polymer with Tg of −36 °C. Copolymers of L-LA and D,L-LA with DXO were studied by Löfgren and Albertsson [129]. The degradation rate in these copolymers can be altered over a broad range. Copolymers of DXO and L-lactic acid containing a high content of lactic acid showed rapid hydrolysis. Crystallinity was observed in these copolymers up to 50% incorporation of DXO [138].

Hydrolytic degradation of poly(ε-CL)/PLLA block copolyester at pH 7.4 and 37 °C over a 5-week period is controlled by the initial crystallinity of the poly(ε-CL) and its overall composition. The rate of degradation increased with increasing PLLA content [203]. Microorganisms, such as *Fusarium solani* and *Fusarium moniliforme*, that secrete poly(ε-CL) depolymerase (cutinase), were more effective with those polymers that had longer poly(ε-CL) sequence lengths [218]. The hydrolytic degradation of the polyester-polyether block copolymer based on poly(ε-CL)/poly(ethylene glycol)/polylactide increased with increasing PEG content. This was attributed to an increased hydrophilicity of the copolymer due to an increase in PEG content [219]. Biodegradation rates of poly(ε-CL-b-PEG) also increased with increasing PEG content [220].

The *in vitro* degradation of poly(trimethylene carbonate-co-caprolactone) containing 80% of ε-CL was similar to that of poly(ε-CL) [221]. The rate of hydrolytic degradation increased in the order of carbonate, ester, and anhydride linkages. This implies that there are differences in the susceptibility of the carbonyl linkage to nucleophilic attack by water molecules. The carbonate is far more resistant to hydrolysis than expected compared to the ester. Degradation times of less than a year can be achieved by using polyesters with a short methylene sequence between the ester bonds like poly(β-PL) and poly(lactide-co-DXO). The degradation rates in many cases are dependent on the initial molecular weight and on the morphology. Electron-beam irradiation and grafting of poly(ε-CL) with acrylamide affected its *in vitro* degradation at pH 7.4 and 37 °C in a phosphate buffer solution [222]. Virgin poly(ε-CL) maintained its Mn and Mw up to 40 weeks, where as a continuous decrease in molecular weight was observed in irradiated (5 Mrad) and grafted polymers.

Blends of poly(ε-CL) with corn starch granules are biodegraded by *Rhizopus arrhizus* lipase and *Bacillus subtilis* α-amylase [188]. A reduction in poly(ε-CL) degradation was observed in its blends with PET [223], whereas in films prepared from its blends with poly(vinyl alcohol) no sign of degradation was observed [202, 224]. Poly(ε-CL) blended with polystyrene, low-density polyethylene, Nylon-6, and PET showed a different susceptibility towards enzymatic degradation than pure poly(ε-CL) [225]. Reactive compatibilized blends of poly(lactic acid) (PLA) and poly(ε-CL), prepared in a twin screw extruder using coupling agents, degraded at a faster rate on enzymatic degradation than the pure poly(ε-CL) or pure PLA. The degradation rates of physical blends were intermediate between those of poly(ε-CL) and PLA [226]. The degradation behavior of solution cast poly(ε-CL) and poly(D,L-lactide) (PDLLA) blend films was studied in phosphate buffer containing *Pseudomonas* lipase (PS). The results indicated the selectivity of PS lipase in promoting degradation of P(ε-CL)[227]. Such behavior was not observed in the presence of porcine pancreatic lipase or *Candida cylindracea* lipase. Biodegradation studies of similar blends in soil have been investigated for 20 months and indicate preferred enzymatic degradation of poly(ε-CL) [228].

The homopolymer of 1,5-dioxepan-2-one (DXO) has good degradation properties. After 46 weeks of *in vitro* degradation, 30–50% of the initial molecular weight is retained [56, 127]. Degradable cross-linked polymers of DXO and a cross-linker; 2,2-bis(ε-CL-4-yl)propane or bis(ε-CL-4-yl), exhibited a degradation time of one year [137]. The amorphous D,L-LA/DXO copolymers degrade faster than the corresponding semicrystalline L-LA/DXO copolymers. An increase in LA content resulted in an increase in degradation rate. The main products of *in vitro* hydrolysis in phosphate buffer (pH 7.4) at 37 °C for a period of up to 20 months of copolymers of DXO and D,L- or L-lactide were identified as the linear monomers, i.e., lactic acid and 2-hydroxyethoxypropionic acid [229]. The amounts of products formed depend upon the copolymer composition and on the degradation time. The *in vitro* degradation studies of PDXO, electron-beam-irradiated PDXO (5 Mrad), and acrylamide-grafted PDXO showed a complete weight loss after 55, 43, and 35 weeks, respectively [149–151].

5
Applications

Aliphatic polyesters are being used in several fields ranging from agricultural implements to biomedical applications. Poly(ε-CL) has been promoted as a soil-degraded container material [230], finding applications for growing and transplanting trees and as a thin-walled tree seedling container [231]. It has also been used for implantable drug delivery devices (Capronol) [232]. Its blends with starch and its derivatives have been used in shopping bags [212, 233].

The potential applications of BIONELLE in injection-molded articles (such as cutlery, brushes, etc.), tubular films (composting bag, shopping bag), flexible packaging, food tray, cosmetic bottles, and beverage bottles have been projected. In some of these applications, there is a need to improve the performance of existing grades.

Poly(ε-CL) is one of the most frequently used compounds in biodegradable drug delivery systems due to its biocompatibilty, low Tg, and high permeability [234]. Blends of PDLLA and PDXO have recently been investigated for the sustained release of a non-steroidal anti-inflammatory drug, diclofenac sodium [179–184]. The release rate is dependent on the blend composition, and this thus provides a powerful means of controlling the drug delivery. Sustained drug delivery has been achieved using both copolymers as well as blends. Morphology is an important factor in determining the *in vitro* performance. Poly(L-LA-co-DXO) exhibited higher moisture sensitivity than a blend of PLLA-PDXO of the corresponding composition, due to its more crystalline and denser morphology. A hydrophilic drug was released much faster from copolymer microspheres than from blend microspheres. Studies on bilayered biodegradable poly(ethylene glycol)/poly(butylene terephthalate) copolymers have recently been reported [234]. The main focus of these studies was to find an optimal polymer matrix for the development of a human skin substitute.

Aliphatic polyesters based on lactic acid have found use in a broad variety of medical applications, bioresorbable surgical sutures, prosthetics, dental implants, pins, bone screws and plates for temporary internal fracture fixation, long-term delivery of antimalarial drugs, contraceptives and eye drugs, and controlled drug delivery. The US Food and Drug Administration has approved only a few biodegradable polymers for use in biological systems [235]. One example is PLGA containing leuprorelin acetate (LH-RH agonist) for the treatment of endometriosis and prostatic cancer. The major application areas of aliphatic polyesters are thus as biomedical polymers and as ecological polymers [236].

A controllable biodegradability, desirable mechanical properties, suitable gas permeability and selectivity would extend the potential application areas of aliphatic polyesters not only in agriculture or in the greenhouse or packaging industry but also as a substitute for human skin. There is a need for such focused studies in the future.

Acknowledgement. The financial assistance provided by The Swedish Research Council to one of the authors (I. K. Varma) is gratefully acknowledged.

References

1. Albertsson A-C, Karlsson S (1996) J Macromol Sci Pure Appl Chem A 33:1565
2. Z.Wirpsza (1993) Polyurethanes, Chemistry, Technology and Applications. Ellis, Horwood, West Sussex
3. Zawadzki SF, Tabak D, Akcelrud L (1993) Polym Plast Technol Engin 32:155
4. Vert M, Li SM, Spenlehaur G, Guerin P (1992) J Mater Sci Mater Med 3:432.
5. Benicewicz BC, Hopper PK (1990) J Bioact Comp Polym 5:453
6. Grijpma DW, Pennings AJ (1994) Macromol Chem Phys 195:1633
7. Holland SJ, Tighe B, Gould PL (1986) J Controlled Release 4:155
8. Lewis DH (1990) In: Chasin M, Langer R (eds), Biodegradable polymers as drug delivery systems. Marcel Dekker, New York, p 1
9. Vert M, Schwarch G, Coudane J (1995) J Macromol Sci Pure Appl Chem A 32:787
10. Mayer JM, Kaplan DL (1994) Trends Polym Sci 2:227
11. Fujimaki T (1998) J Polym Degrad Stabil 59:209
12. Amass W, Amass A, Tighe B (1998) Polym Int 47:89
13. Albertsson A-C (1992) In: Hamid SH, Maadhah AG, Amin MB (eds), Handbook of polymer degradation. Marcel Dekker, New York, p 345
14. Albertsson A-C, Karlsson S (1992) In: Allen G, Aggarwal SL, Russo R (eds), Comprehensive polymer science, First supplement. Pergamon Press, Oxford, p 285
15. Holland SJ, Tighe BJ (1992) In: Advances in Pharmaceutical Sciences. Vol 6, Academic, San Diego, CA, Chap 4, p 101
16. Löfgren A, Albertsson A-C, Dubois P, Jérôme R (1995) J Macromol Sci Rev Macromol Chem Phys C 35:379
17. Albertsson A-C, Karlsson S (1996) In: Salamone JC (ed), Polymeric materials encyclopedia; Synthesis, properties and applications. CRC Press Inc, Boca Raton, p 150
18. Albertsson A-C (1999) In: Hamid SH, Maadhah AG, Amin MB (eds), Handbook of polymer degradation. Marcel Dekker, New York, p 417
19. Albertsson A-C, Karlsson S (1998) Biodegradable Polymers. In: Meyers RA (ed), Encyclopedia of Environmental Analysis and Remediation. John Wiley and Son, New York, p 641
20. Arvanitoyannis IS (1999) J Macromol Sci Revs Macromol Chem Phys C 39:205
21. Albertsson A-C, Varma IK (2001) In: Doi Y, Steinbüchel A (eds), Biopolymers. Wiley-VCH, Weinheim, in press
22. Ikada Y, Tsuji H (2000) Macromol Rapid Commun 21:117
23. Deckwer WD (1995) FEMS Microbiol Rev 16:143
24. Nghiem NP, Davison BH, Suttle BE, Richardson GR (1997) Appl Biochem Biotech 63–65:565
25. Lee PC, Lee WG, Lee SY, Chang HN (1999) Process Biochem 35:49
26. Varma DS, Maheshwari A, Gupta V, Varma IK (1980) Angew Makromol Chem 90:23
27. Varma DS, Negi YS, Veena, Varma IK (1983) ACS Polym Prepr Div Polym 24:245
28. Varma DS, Agarwal R, Varma IK (1985) Brit Polym J 17:83

29. Varma DS, Agarwal R, Varma IK (1985) Polym Commun 26:346
30. Varma DS, Agarwal R, Varma IK (1985) Polym Mater Sci & Engg 53:689
31. Varma IK, Negi YS, Varma DS (1986) J Polym Mater 3:197
32. Varma DS, Negi YS, Choudhary V, Varma IK (1986) Ind J Text Res 11:173
33. Reed AM, Gilding DK (1981) Polymer 22:499
34. Gilding DK, Reed AM (1979) Polymer 20:1454
35. Witt U, Muller R-J, Augusta J, Widdecke H, Deckwer W-D (1994) Macromol Chem Phys 195:793
36. Jun HS, Kim BO, Kim YC, Chang HN, Woo SI (1994) J Environ Polym Degrad 2:9
37. Witt U, Muller R-J, Deckwer W-D (1995) J Environ Polym Degrad 3:215
38. Casey DJ, Roby MS (1984) Eur Pat Appl (Am Cyanamid Co) 108,933 (cf CA 1984, 101:920841)
39. Albertsson A-C, Ljungquist O (1986) J Macromol Sci Chem A 23:393
40. Kobayashi T, Hori Y, Kakimoto MA, Imai Y (1993) Makromol Chem Rapid Commun 14:785
41. Wang HH, Chen JC (1995) Polym Eng Sci 35:1468
42. Albertsson A-C, Ljungquist O (1986) J Macromol Sci Pure Appl Chem A 23:411
43. Albertsson A-C, Ljungquist O (1987) J Macromol Sci Pure Appl Chem A 24:977
44. Ranucci E, Liu Y, Söderqvist Lindblad M, Albertsson A-C (2000) Macromol Rapid Commun 21:680
45. Knani D, Gutman AL, Kohn D (1993) J Polym Sci Part A Polym Chem 31:1221
46. Knani D, Kohn D (1993) J Polym Sci Part A Polym Chem 31:2887
47. Athawale VD, Gaonkar SR (1994) Biotechnol Lett 16(2):149
48. Mezoul G, Lalot T, Brigodiot M, Marechal E (1995) J Polym Sci Part A Polym Chem 33:2691
49. Brazwell EM, Filos D, Morrow CJ (1995) J Polym Sci Part A Polym Chem 33:89
50. Suda S, Uyama H, Kobayashi S (1999) Proceed Japan Academy Ser B Physical Biol Sci 75:201
51. Kobayashi S, Uyama H, Namekawa S (1998) Polym Degrad Stab 59:195
52. Shuai XT, Jedlinski Z, Kowalczuk M, Rydz J, Tan HM (1999) Eur Polym J 35:721
53. Kafrawy A, Mattei FV, Shalaby SW (1984) US Patent 4,470,416 (to Ithicon Inc)
54. Claridge DV (1972) British Patent 1,272,733
55. Mathisen T, Albertsson A-C (1989) Macromolecules 22:3838
56. Mathisen T, Masus K, Albertsson A-C (1989) Macromolecules 22:3842
57. Kafrawy A, Shalaby SW (1987) J Polym Sci Polym Chem Ed 25:2629
58. Bailey WJ (1985) Polym J 17:85
59. Bailey WJ (1989) In: Allen G, Bevington JC (eds), Comprehensive polymer science, The synthesis, characterisation and applications of polymers. Pergamon Press, Oxford, vol. 3, p 283
60. Duda A, Biela T, Libiszowski J, Penczek S, Dubois P, Mecerreyes D, Jérôme R (1998) Polym Degrad Stab 59:215
61. Duda A, Penczek S (1990) Macromolecules 23:1636
62. Evstropov A, Lebdev BV, Kiparisova, EG, Alekseev VA, Stashina, GA (1980) Vysokomol Soedin Ser A 22:2450
63. Slomkowski S, Sosnowski, Gadzinowski M (1998) Polym Degrad Stabil 59:153
64. Mecerreyes D, Jérôme R, Dubois P (1999) Adv Polym Sci 147:1
65. Mecerreyes D, Jérôme R (1999) Macromol Chem Phys 200:2581
66. Albertsson A-C, Löfgren A (1996) In: Salamone JC (ed), Polymeric materials encyclopedia: Synthesis, properties and applications. CRC Press Inc, Boca Raton, p 634
67. Fukuzaki H, Yoshida M, Asano M, Kumakura M (1989) Eur Polym J 25(10):1019
68. Fukuzaki H, Aika Y, Yoshida M, Asano M, Kumakura M (1989) Makromol Chem 190:1553
69. Fukuzaki H, Yoshida M, Asano M, Aika Y, Kumakura M (1989) Eur Polym J 26(4):457
70. Cerrai P, Tricoli M, Andruzzi F, Paci M (1989) Polymer 30:338

71. Dong H, Cao S-G, Li Z-Q, Han S-P, You D-L, Shen J-C (1999) J Polym Sci Part A Polym Chem 37:1265
72. Uyama H, Kobayashi S (1993) Chem Lett 1149
73. Uyama H, Takeya K, Hoshi N, Kobayashi S (1995) Macromolecules 28:7046
74. Svirkin YY, Xu J, Gross RA, Kaplan DL, Swift G (1996) Macromolecules 29:4591
75. MacDonald RT, Pulapura SK, Svirkin Y, Gross RA, Kaplan DL, Akkara J, Swift G, Wolk S (1995) Macromolecules 28:73
76. Henderson LA, Svirkin YY, Gross RA, Kaplan DL, Swift G (1997) Macromolecules 30:7759
77. Bisht KS, Henderson LA, Gross RA, Kaplan DL, Swift G (1997) Macromolecules 30:2705
78. Nobes GAR, Kazlauskas RJ, Marchessault RH (1996) Macromolecules 29:4829
79. Bailey WJ, Ou JM, Zhou L-L (1990) ACS Polym Prepr Div Polym Chem 31(1):24
80. Tadokoro A, Takata T, Endo T (1993) Macromolecules 26:4400
81. Kurcok P, Kowalczuk M, Hennek K, Jedlinski Z (1992) Macromolecules 25:2017
82. Kurcok P, Penczek J, Franek J, Jedlinski Z (1992) Macromolecules 25:2285
83. Jedlinski Z, Kurock P, Kowalczuk M (1985) Macromolecules 18:2679
84. Jedlinski Z, Kowalczuk M (1989) Macromolecules 22:3242
85. Jedlinski Z, Kowalczuk M, Kurcok P (1991) Macromolecules 24:1218
86. Kricheldorf HR, Boettcher C (1993) Makromol Chem 194:1665; Makromol Chem Macromol Symp 73:47
87. Jedlinski Z, Kurcok P, Walach W, Janeczek H Radecka I (1993) Makromol Chem 194:1681
88. Albertsson A-C, Löfgren A, Sjöling M (1993) Makromol Chem Macromol Symp 73:127
89. Okamoto Y (1990) ACS Polym Prepr Div Polym Chem 31 (1):10
90. Okamoto Y (1991) Makromol Chem Macromol Symp 42/43:117
91. Albertsson A-C, Palmgren R (1996) J Macromol Sci Pure Appl Chem A 33:747
92. Kricheldorf HR, Berl M, Scharnagl N (1988) Macromolecules 21:286
93. Kricheldorf HR, Sumbel MV, Keiser-Saunders I (1991) Macromolecules 24:1944
94. Hofman A, Slomkowski S, Penczek S (1987) Makromol Chem Rapid Commun 8:387
95. Löfgren A, Albertsson A-C, Dubois P, Jérôme R, Teyssié P (1994) Macromolecules 27:5556
96. Dubois P, Jérôme R, Teyssié P (1991) Makromol Chem Macromol Symp 42/43:103
97. Dubois P, Jacobs C, Jérôme R, Teyssié P (1991) Macromolecules 24:2266
98. Vanhoorne P, Dubois P, Jérôme R, Teyssié P (1992) Macromolecules 25:37
99. Inoue S (1988) J Macromol Sci Pure Appl Chem A 25:571
100. Shimasaki K, Aida T, Inoue S (1987) Macromolecules 20:3076
101. McLain SJ, Drysdale NE (1992) Polym Prepr (Am Chem Soc Div Polym Chem) 33(1):174
102. McLain SJ, Ford TM, Drysdale NE (1992) Polym Prepr (Am Chem Soc Div Polym Chem) 33(2):463
103. Shen ZQ, Shen YQ, Sun JQ, Zhang FY, Zhang YF (1994) Chin Sci Bull 39:1096
104. Shen YQ, Shen ZQ, Zhang FY, Zhang YF (1995) Polym J 27:59
105. Stevels WM, Ankoné MJK, Dijkstra PJ, Feijen J (1996) Macromolecules 29:8296
106. Stevels WM, Ankoné MJK, Dijkstra PJ, Feijen J(1996) Macromolecules 29:3332
107. Shen Y, Shen Z, Shen J, Zhang Y, Yao K (1996) Macromolecules 29:3441
108. Shen Y, Shen Z, Zhang Y, Yao K (1996) Macromolecules 29:8289
109. Shen Y, Zhu KJ, Shen Z, Yao K (1996) J Polym Sci Part A Polym Chem 34:1799
110. Agarwal S, Mast C, Dehnicke K, Grenier A (2000) Macromol Rapid Commun 21:195
111. Dubois P, Degee P, Ropsen N, Jérôme R (1997) In: Hatada K, Kitayama T, Vogl O (eds), Macromolecular design of polymeric materials. Marcel Dekker, New York, p 247
112. Chamberlain BM, Sun Y, Hagadorn JR, Hemmesch EW, Young Jr VG, Pink M, Hillmyer MA, Tolman WB (1999) Macromolecules 32:2400
113. Baran J, Duda A, Kowalski A, Szymanski R, Penczek S (1998) Macromol Symp 128:241
114. Lundmark S, Sjöling M, Albertsson A-C, (1991) J Macromol Sci Chem A 28:15

115. Inoue S, Aida T (1993) Makromol Chem Macromol Symp 73:27
116. Abraham GA, Gallardo A, Lozano AE, Roman JS (2000) J Polym Sci Part A Polym Chem 38:1355
117. Rafler G, Dahlmann J (1992) Acta Polym 43:91
118. Nijenhuis AJ, Grijpma DW, Pennings AJ (1992) Macromolecules 25:6419
119. Kowalski A, Duda A, Penczek S (2000) Macromolecules 33:689
120. Kowalski A, Duda A, Penczek S (1998) Macromol Rapid Commun 19:567
121. Kowalski A, Libiszowski J, Duda A, Penczek S (1998) ACS Polym Prepr Div Polym Chem 39(2):74
122. Kricheldorf HR, Kreiser-Saunders I, Stricker A (2000) Macromolecules 33:702
123. Kricheldorf HR, Kreiser-Saunders I (2000) Polymer 41:3957
124. Kricheldorf HR, Eggerstedt S (1998) Macromol Chem Phys 199:283
125. Mathisen T, Albertsson A-C (1990) J Appl Polym Sci 38:591
126. Albertsson A-C, Löfgren A (1992) Makromol Chem Macromol Symp 53:221
127. Albertsson A-C, Palmgren R (1993) J Macromol Sci Pure Chem A 30:919
128. Albertsson A-C, Eklund M (1994) J Polym Sci Part A Polym Chem 32:265
129. Löfgren A, Albertsson A-C (1994) J Appl Polym Sci 52:1327
130. Albertsson A-C, Löfgren A, Zhang Y, Bjursten L-M (1994) J Biomater Sci Polym Ed 6:411
131. Stridsberg K, Gruvegård M, Albertsson A-C (1998) Macromol Symp 130:367
132. Albertsson A-C, Palmgren R (1994) Macromol Reports A 31:1185
133. Albertsson A-C, Löfgren A (1995) J Macromol Sci Pure Appl Chem A 32:41
134. Albertsson A-C, Gruvegård M (1995) Polymer 36:1009
135. Löfgren A, Renstad R, Albertsson A-C (1995) J Appl Polym Sci 55:1589
136. Löfgren A, Albertsson A-C (1995) Polymer 36:3753
137. Palmgren R, Karlsson S, Albertsson A-C (1997) J Polym Sci Part A Polym Chem 35:1635
138. Gruvegård M, Lindberg T, Albertsson A-C (1998) J Macromol Sci Pure Appl Chem A 35(6):885
139. Liu Y, Schulze M, Albertsson A-C (1998) J Macromol Sci Pure Appl Chem A 35(2):207
140. Penczek S, Duda A (1996) Macromol Symp 107:1; Kricheldorf HR, Serra A (1985) Polymer Bull 14:497
141. Möller M, Kånge R, Hedrick JL (2000) J Polym Sci Part A Polym Chem 38:2067
142. Nomura N, Taira A, Tomioka T, Okada M (2000) Macromolecules 33:1497
143. Agarwal S, Mast C, Anfang S, Carl M, Dehnicke K, Greiner A (1998) ACS Polym Prepr Div Polym Chem 39:414
144. Agarwal S, Karl M, Dehnicke K, Grenier A (1998) ACS Polym Prepr Div Polym Chem 39:361
145. Agarwal S, Brandukova-Szmikowski NE, Greiner A (1999) Macromol Rapid Commun, 20:274
146. Agarwal S, Brandukova-Szmikowski NE, Grenier A (1999) Polym Adv Technol 10:528
147. Pilati F, Toselli M, Messori M, Priola A, Bongiovanni R, Malucelli G, Tonelli C (1999) Macromolecules 32:6969
148. Stridsberg K, Albertsson A-C (2000) J Polym Sci Part A Polym Chem 38:1774
149. Ohrlander M, Wirsén A, Albertsson A-C (1999) J Polym Sci Part A Polym Chem 37:1643
150. Ohrlander M, Palmgren R, Wirsén A, Albertsson A-C (1999) J Polym Sci Part A Polym Chem 37:1659
151. Ohrlander M, Lindberg T, Wirsén A, Albertsson A-C (1999) J Polym Sci Part A Polym Chem 37:1651
152. Mecerreyes D, Miller RD, Hedrick JL, Detrembleur C, Jérôme R (2000) J Polym Sci Part A Polym Chem 38:870
153. Detrembleur C, Mazza M, Halleux O, Lecomte P, Mecerreyes D, Hedrick JL, Jérôme R (2000) Macromolecules 33:14
154. Tian D, Dubois P, Grandfils C, Jérôme J (1997) Macromolecules 30:406

155. Tian D, Halleux O, Dubois P, Jérôme R, Sobry R, van den Bossche G (1998) Macromolecules 31:924
156. Shi W, Rånby B (1996) J Appl Polym Sci 59:1937
157. Shi W, Rånby B (1996) J Appl Polym Sci 59:1945
158. Shi W, Rånby B (1996) J Appl Polym Sci 59:1951
159. Hult A, Johansson M, Malmström E (1999) Adv Polym Sci 143:1
160. Trollsås M, Hedrick JL (1998) J Am Chem Soc 120:4644
161. Trollsås M, Hedrick JL (1998) Macromolecules 31:4390
162. Trollsås M, Hedrick JL, Mecerreyes D, Jérôme R, Dubois P (1998) J Polym Sci Part A Polym Chem 36:3187
163. Trollsås M, Hawker CJ, Remenar JF, Hedrick JL, Johansson M, Ihre H, Hult A (1998) J Polym Sci Part A Polym Chem 36:2793
164. Trollsås M, Atthoff B, Claesson H, Hedrick JL (1998) Macromolecules 31:3439
165. Hawker CJ, Frechet JMJ (1990) J Am Chem Soc 112:7638
166. Trollsås M, Löwenhielm P, Lee VY, Möller M, Miller RD, Hedrick JL (1999) Macromolecules 32:9062
167. Trollsås M, Kelly MA, Claesson H, Siemens R, Hedrick JL (1999) Macromolecules 32:4917
168. Liu MJ, Vladimorov N, Frechet JMJ (1999) Macromolecules 32:6881
169. Cordova A, Hult A, Hult K, Ihre H, Iversen T, Malmstrom E (1998) J Am Chem Soc 120:13521
170. Eastmond GC (1999) Adv Polym Sci 149:59
171. Grijpma DW, Joziasse C, Pennings A (1993) Makromol Chem Rapid Commun 14:155
172. Jin H-J, Park J-K, Park K-H, Kim M-N, Yoon J-S (2000) J Appl Polym Sci 77:547
173. Vert M (1984) Angew Makromol Chem 166/167:155
174. Aubin M, Prud'homme RE (1981) Polymer 22:1223
175. Mathisen T, Lewis M, Albertsson A-C (1991) J Appl Polym Sci 42:2365
176. Chen H-L, Li L-J, Ou-Yang WC, Hwang JC, Wong W-Y (1997) Macromolecules 30:1718
177. Bastioli C, Cerutti A, Guanella I, Romano GC, Tosin M (1995) J Environ Polym Degrad 3:81
178. Albertsson A-C, Löfgren A (1995) J Macromol Sci Pure Appl Chem A 32:41
179. Edlund U, Albertsson A-C (2000) J Polym Sci Part A Polym Chem 38:786
180. Ajji A, Renand MC (1991) J Appl Polym Sci 26:3917
181. Chiu S-C, Smith TG (1984) J Appl Polym Sci 29:1797
182. Edlund U, Albertsson A-C (1999) J Polym Sci Part A Polym Chem 37:1877
183. Edlund U, Albertsson A-C (2000) J Bioact Compat Polym 15:1
184. Wang S, Nishide H, Tsuchida E (1999) Polym Advanced Tech 10:282
185. Albertsson, A-C (1993) J Macromol Sci Pure Appl Chem A 30:757
186. Karlsson S, Albertsson A-C (1995) J Macromol Sci Pure Appl Chem A 32:599
187. Takiyama E, Fujimaki T (1992) Plastics 43:87
188. Tokiwa Y, Iwamoto A, Koyama M (1990) Polym Mater Sci Eng 63:742
189. Mochizuki M, Hirami M (1997) Polym Advanced Tech 8:203
190. Park TG, Cohen S, Langer R (1992) Macromolecules 25:116
191. Zhu KJ, Hendren RW, Jensen K, Pitt CG (1991) Macromolecules 24:1736
192. Reeve MS, McCarthy SP, Downey MJ, Gross RA (1994) Macromolecules 27:825
193. Williams DF (1992) Clin Mater 10:9
194. Albertsson A-C, Ljungquist O (1988) J Macromol Sci Pure Appl Chem A 25:467
195. Kasuya K, Takagi K, Ishiwatari S, Yoshida Y, Doi Y (1998) Polym Degrad Stab 59:327
196. Khan MA, Idriss Ali KM, Yoshii F, Makuuchi K (1999) Polym Degrad Stab 63:261
197. Nagata M, Kono Y, Sakai W, Tsutsumi N (1999) Macromolecules 32:7762
198. Jang JP, Lee KH, Kim MN (1996) Polym Advanced Tech 8:146
199. Goldberg D (1995) J Environ Polym Degrad 3:61
200. Albertsson A-C, Renstad R, Erlandsson B, Eldsäter C, Karlsson S (1998) J Appl Polym Sci 70:61

201. Pettigrew CA, Reece GA, Smith MC, King LW (1995) J Macromol Sci Pure Appl Chem A 32:811
202. Yasin M, Tighe B (1992) Biomaterials 13:9
203. Ye WP, Du FS, Zin WH, Yang Y, Xu Y (1997) React Funct Polym 32:161
204. Fields RD, Rodriguez F, Finn RK (1974) J Appl Polym Sci 18:3571
205. Tokiwa Y, Endo T, Suzuki T (1976) J Ferment Tech 54:603
206. Tokiwa Y, Suzuki T (1977) Nature 270:76
207. Tilstra L, Johnsonbaugh D (1993) J Environ Polym Degrad 1:247
208. Eldsäter C, Erlandsson B, Renstad R, Albertsson A-C, Karlsson S (2000) Polymer 41:1297
209. Karlsson S, Albertsson A-C (1998) Polym Eng Sci 38:1251
210. Karlsson S, Albertsson A-C (1998) Macromol Symp 127:219
211. Chen H, Bei J-Z, Wang S (2000) Polym Advanced Tech 11:180
212. Darwis D, Mitomo H, Enjoji T, Yoshii F, Makuuchi K (1998) Polym Degrad Stab 62:259
213. Nishida H, Tokiwa Y (1993) J Environ Polym Degrad 1:227
214. Mochizuki M, Hirano M, Kanmuri Y, Kudo K, Tokiwa Y (1995) J Appl Polym Sci 55:289
215. Murphy CA, Cameron JA, Huang SJ, Vinopal RT (1996) Appl Environ Microbiol 62:456
216. Oda Y, Asari H, Urakami T, Tonomura K (1995) J Ferment Bioeng 80:265
217. Schindler A, Jeffcoat R, Kimmel GL, Pitt CG, Wall ME, Zweidinger R (1977) In: Pearce EM, Schaefgen JR (eds), Contemporary Topics in Polymer Science. Plenum, New York, Vol 2, p 251
218. Lostocco MR, Murphy CA, Cameron JA, Huang SJ (1998) Polym Degrad Stab 59:303
219. Chen DR, Bei JZ, Wang SG (2000) Polym Degrad Stab 67:455
220. Bei J-Z, Li J-M, Wang Z-F, Le J-C, Wang S-G (1996) Polym Advanced Technol 8:693
221. Albertsson A-C, Eklund M (1995) J Appl Polym Sci 57:87
222. Ohrlander M, Erickson R, Palmgren R, Wirsén A, Albertsson A-C (2000) Polymer 41:1277
223. Chiellini E, Corti A, Giovannini A, Narducci P, Paparella AM, Solaro R (1996) J Environ Polym Degrad 4:37
224. de Kesel C, Wauven CV, David C (1997) Polym Degrad Stab 55:107
225. Albertsson A-C, Karlsson S (1995) Acta Polym 46:114
226. Wang LW, Ma W, Gross RA, McCarthy SP (1998) Polym Degrad Stab 59:161
227. Gan Z, Yu D, Zhong Z, Liang Q, Jing X (1999) Polymer 40:2859
228. Tsuji H, Mizuno A, Ikada Y (1998) J Appl Polym Sci 70:2259
229. Karlsson S, Hakkarainen M, Albertsson A-C (1994) J Chromatogr A 688:251
230. Pitt CG, Chasalow FI, Hibionada YM, Klimas DM, Schindler A (1981) J Appl Polym Sci 26:3779
231. Raghavan D (1995) Polym-Plast Technol Eng 34:41
232. Engelberg I, Kohn J (1991) Biomaterials 12:292
233. Ali SAM, Zhong S-P, Doherty PJ, Williams DF (1993) Biomaterials 14:648
234. van Dorp AGM, Verhoeven MCH, Koerten HK, van Blitterswijk CA, Ponec M (1999) J Biomed Mater Res 47:292
235. Davis S, Illum L, Stolnik S (1996) Curr Opin Colloid Interface Sci 660
236. Ikada Y, Tsuji H (2000) Macromol Rapid Commun 2:117

Received: January 2001

Controlled Ring-Opening Polymerization: Polymers with designed Macromolecular Architecture

Kajsa M. Stridsberg, Maria Ryner, Ann-Christine Albertsson

Department of Polymer Technology, The Royal Institute of Technology, 100 44 Stockholm, Sweden
e-mail: aila@polymer.kth.se

Abstract. This paper reviews ring-opening polymerization of lactones and lactides with different types of initiators and catalysts as well as their use in the synthesis of macromolecules with advanced architecture. The purpose of this paper is to review the latest developments within the coordination-insertion mechanism, and to describe the mechanisms and typical kinetic features. Cationic and anionic ring-opening polymerizations are mentioned only briefly.

Keywords. Ring-opening polymerization, Lactones, Lactides, Kinetics, Coordination-insertion mechanism, Macromolecular architecture

1	Introduction	42
1.1	Ring-Opening Polymerization	43
1.2	Ring-Opening Polymerization of Cyclic Esters	43
1.2.1	Cationic Ring-Opening Polymerization	44
1.2.2	Anionic Ring-Opening Polymerization	45
1.2.3	"Coordination-Insertion" Ring-Opening Polymerization	45
2	Initiators	46
2.1	Initiators for the ROP of Lactones and Lactides	46
2.1.1	Transesterification Reactions	46
2.1.2	Tin(II) 2-Ethylhexanoate	48
2.1.3	Aluminum Tri-Isopropoxide	49
2.1.4	Tin(IV) Alkoxides	50
2.1.5	Tin(II) Alkoxides	52
2.1.6	Lanthanide Alkoxides	52
3	Kinetics of Ring-Opening Polymerization	52
3.1	Kinetic Models	52
3.2	Kinetics of Lactide Polymerization	53

4	Macromolecular Architecture	55
4.1	Homopolymers	55
4.2	Block Copolymers	55
4.3	Star-Shaped (Co)Polymers	57
5	**Biodegradable Polymers**	58
5.1	Polymer Degradation	58
5.1.1	Polyglycolide and Copolymers	59
5.1.2	Polylactide and Copolymers	59
5.1.3	Poly(ε-caprolactone) and Copolymers	61
6	**Conclusions**	61
References		62

Abbreviations

ε-CL	ε-caprolactone
D-LA	D,L-lactide
DXO	1,5-dioxepan-2-one
FDA	Food and Drug Administration
I	initiator
L-LA	L,L-lactide
M	monomer
MWD	molecular weight distribution
P	polymer
rac	racemic
ROP	ring-opening polymerization
$Sn(Oct)_2$	tin(II) 2-ethylhexanoate
THF	tetrahydrofuran

1
Introduction

Aliphatic polyesters are an attractive class of polymer that can be used in biomedical and pharmaceutical applications. One reason for the growing interest in this type of degradable polymer is that their physical and chemical properties can be varied over a wide range by, e.g., copolymerization and advanced macromolecular architecture. The synthesis of novel polymer structures through ring-opening polymerization has been studied for a number of years [1–5]. The development of macromolecules with strictly defined structures and properties, aimed at biomedical applications, leads to complex and advanced architecture and a diversification of the hydrolyzable polymers.

Degradable materials with new mechanical properties and modified degradation profiles have been produced and characterized. The increasing demands of a larger number of biomedical applications have resulted in an increasing interest in producing macromolecules through controlled polymerization.

1.1
Ring-Opening Polymerization

Polylactones and polylactides can be prepared by two different approaches, by the polycondensation of hydroxycarboxylic acids or by the ring-opening polymerization (ROP) of cyclic esters. The polycondensation technique is less expensive than ROP, but it is difficult to obtain high molecular weight polymers, to achieve specific end groups, and to prepare well-defined copolyesters. The ROP of lactones and lactides has been thoroughly investigated during the last 40 years, due to its versatility in producing a variety of biomedical polymers in a controlled manner. Carothers and coworkers first extensively explored the ROP technique for lactones, anhydrides, and carbonates [6–9]. Since then the method has been applied to a diversity of monomers to produce all types of polymers, and a number of initiator and catalyst systems have been developed. In many cases, the resulting polymers exhibit useful properties as engineering materials.

There are several reasons for studying the polymerization of cyclic esters. First, to exploit the potential of synthetic polymer chemistry to prepare a variety of polymers with control of the major variables affecting polymer properties. Experimental conditions have to be optimized in order to find the best polymerization system for a desired technological or industrial process. Factors such as economy, toxicology, and technical apparatus development are important. A second reason for studying ROP is to enable various advanced macromolecules, including homopolymers with well-defined structures or end groups, to be prepared, as well as copolymers with different architectures, e.g., block, graft, or star copolymers. The physical, mechanical, and degradation properties of these various macromolecules are studied to determine the structure-to-property relationship. The third reason for studying these kinds of systems is that they are valuable models for the examination of the kinetics [10] and mechanisms [11] of elementary reactions in polymerization.

1.2
Ring-Opening Polymerization of Cyclic Esters

Polylactones and polylactides of high molecular weight are exclusively produced by the ROP of the corresponding cyclic monomers. A polyester is formed when cyclic esters are reacted with a catalyst or initiator. Scheme 1 presents the reaction pathway for the ROP of a cyclic ester.

Each macromolecule formed generally contains one chain end terminated with a functional group originating from the termination reaction and one terminus end capped with a functional group originating from the initiator. By al-

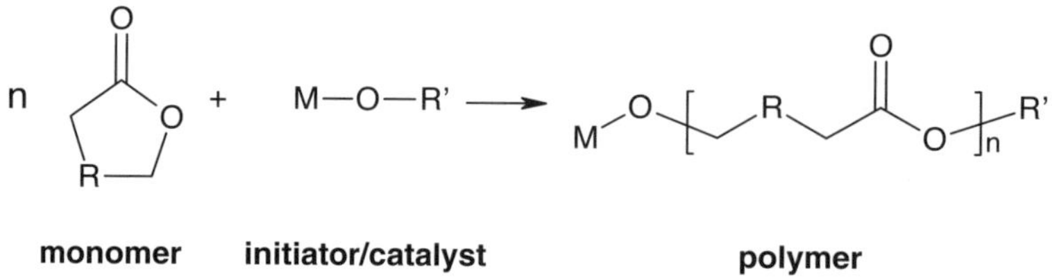

Scheme 1. Schematic representation of the ROP of a cyclic ester. R=$(CH_2)_{0-3}$ and/or (CHR")

tering the catalyst or initiator and the termination reaction, the nature of the functional groups can be varied to fit the application of the polymer. The types of initiator and end group play important roles in determining both the thermal stability and hydrolytic stability of the resulting polyester [12–14]. Functional groups accessible to post-polymerization reactions can also be introduced into the polymer structure in this way.

The ring-opening reaction can be performed either as a bulk polymerization, or in solution, emulsion, or dispersion [15, 16]. A catalyst or initiator is necessary to start the polymerization. Under rather mild conditions, high-molecular weight aliphatic polyesters of low polydispersity can be prepared in short periods of time. Problems associated with condensation polymerization, such as the need for exact stoichiometry, high reaction temperatures, and the removal of low molecular weight by-products (e.g., water) are excluded in ROP [17].

Depending on the initiator, the polymerization proceeds according to three different major reaction mechanisms [18], *viz.*, cationic, anionic, or "coordination-insertion" mechanisms [19–21]. In addition, radical, zwitterionic [22], or active hydrogen[18] initiation is possible, although such techniques are not used to any great extent. The focus in this review is on the "coordination-insertion" mechanism and the other methods are described only briefly.

1.2.1
Cationic Ring-Opening Polymerization

Among the cyclic esters, 4-, 6-, and 7-membered rings form polyesters when reacted with cationic catalysts [18, 23–25]. The cationic ROP involves the formation of a positively charged species which is subsequently attacked by a monomer (Scheme 2). The attack results in a ring-opening of the positively charged species through an S_N2-type process.

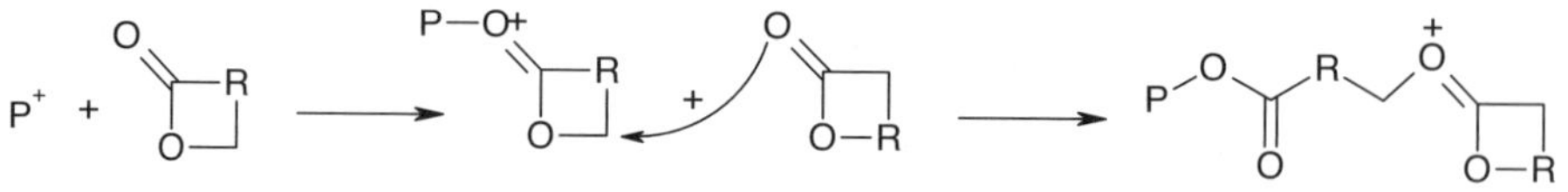

Scheme 2. The reaction pathway for the ROP of a cyclic ester by cationic initiation

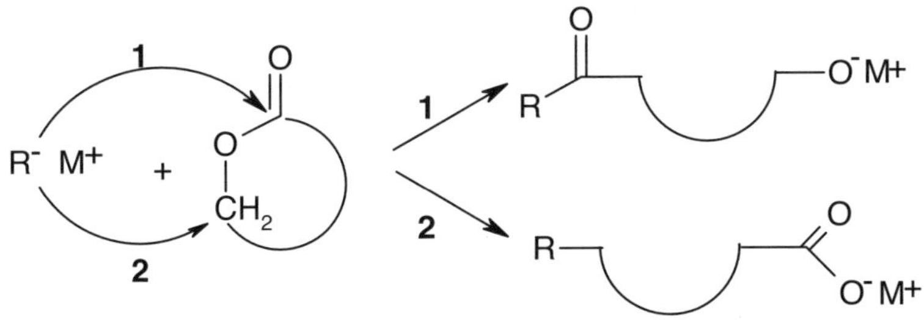

Scheme 3. The reaction pathway for the ROP of a cyclic ester by anionic initiation. Ring-opening of monomer by 1) acyl-oxygen bond cleavage and 2) alkyl-oxygen bond cleavage

The cationic polymerization is difficult to control and often only low-molecular weight polymers are formed. When the bulk and solution polymerization of 1,5-dioxepan-2-one (DXO) with cationic initiators were studied, the highest molecular weight achieved was about 10,000 [23]. More detailed reviews on cationic ROP have been published by Penczek and coworkers [26, 27].

1.2.2
Anionic Ring-Opening Polymerization

Anionic ROP of cyclic ester monomers takes place by the nucleophilic attack of a negatively charged initiator on the carbonyl carbon or on the carbon atom adjacent to the acyl oxygen, resulting in a linear polyester (Scheme 3) [28, 29] The propagating species is negatively charged and is counter-balanced with a positive ion. Depending on the nature of the ionic propagating chain end and the solvent, the reacting complex varies from completely ionic to almost covalent.

One of the best controlled methods leading to high molecular weight polymers is anionic polymerization carried out in a polar solvent. The Jedlinski group developed living anionic ROP methods for 4- and 5-membered ring lactones and has reported well-defined polymers and copolymers of high molecular weight [30]. The anionic ring-opening of four-membered rings (β-lactones) occurs through alkyl-oxygen or acyl-oxygen cleavage giving a carboxylate or alkoxide [31]. Larger lactones, such as ε-caprolactone (ε-CL) or lactide, react only by an attack of the anion on the carbonyl carbon atom with acyl-oxygen scission and the formation of an alkoxide as the growing species [32,33]. A problem associated with the anionic ROP is the extensive back-biting, and in some cases only polyesters of low molecular weight are achieved [34].

1.2.3
Coordination-Insertion Ring-Opening Polymerization

The pseudo-anionic ROP is often referred to as coordination-insertion ROP, since the propagation is thought to proceed by coordination of the monomer to the active species, followed by insertion of the monomer into the metal-oxygen

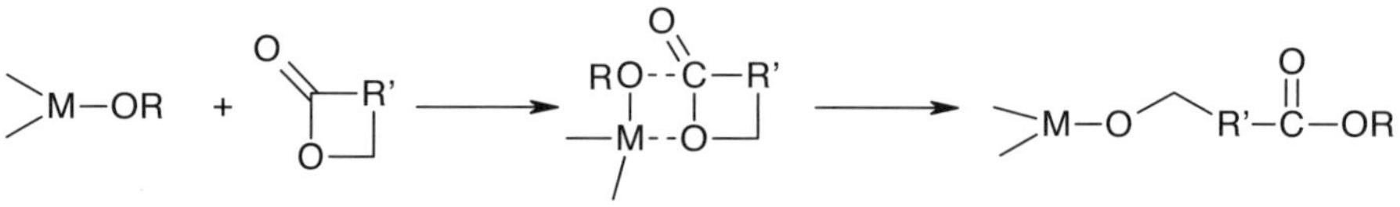

Scheme 4. The proposed reaction pathway for the ROP of a cyclic ester by the coordination-insertion mechanism

bond by rearrangement of the electrons [19, 20]. Scheme 4 shows a schematic presentation of the coordination-insertion mechanism. The growing chain remains attached to the metal through an alkoxide bond during the propagation. The reaction is terminated by hydrolysis forming a hydroxy end group. With functional alkoxy-substituted initiators, macromers with end groups active in post-polymerization reactions are produced.

The coordination-insertion type of polymerization has been thoroughly investigated since it may yield well-defined polyesters through living polymerization [20]. When two monomers of similar reactivity are used, block copolymers can be formed by sequential addition to the "living" system [63].

2
Initiators

2.1
Initiators for the ROP of Lactones and Lactides

The synthesis of novel initiators and the ROP of existing or new monomers and macromers substituted with functional groups provide a very interesting and promising strategy for producing structurally advanced macromolecules.

A large variety of organometallic compounds, e.g., metal alkoxides and metal carboxylates, has been studied as initiators or catalysts in order to achieve effective polymer synthesis [35]. Many reactions catalyzed by metal complexes are highly specific and, by careful selection of metal and ligands, reactions can be generated to form a desired polymer structure [36, 37]. The covalent metal alkoxides with free p or d orbitals react as coordination initiators and not as anionic or cationic inititors [38]. Fig. 1 summarizes some of the most frequently used initiators and catalysts.

2.1.1
Transesterification Reactions

It is well known from the ROP of lactones and lactides that the catalyst or initiator causes transesterification reactions at elevated temperatures [39], or at long reaction times (Scheme 5) [40]. Intermolecular transesterification reactions modify the sequences of copolylactones and prevent the formation of block copolymers. Intramolecular transesterification reactions, i.e., back-biting, cause

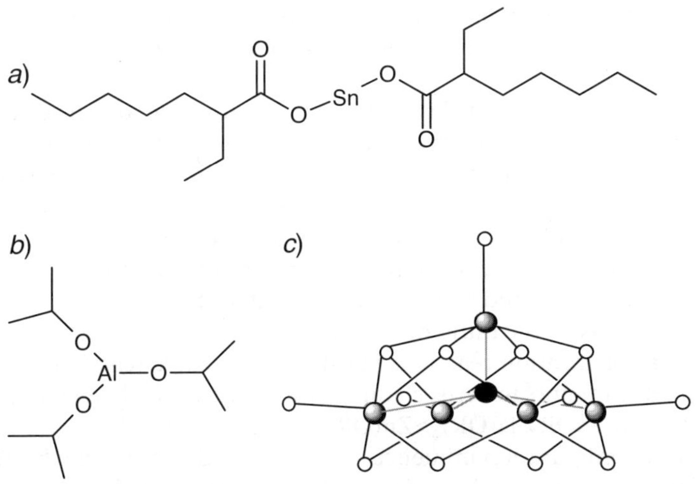

Fig. 1. Chemical structure of initiators used in ROP of lactones and lactides. *a*) stannous octoate, *b*) aluminum isopropoxide, *c*) lanthanide isopropoxide. Lanthanum atoms are represented by gray circles and oxygen atoms by white circles. The black circle represents the bridging oxygen atom connecting all lanthanum atoms. Alkyl groups are omitted for clarity

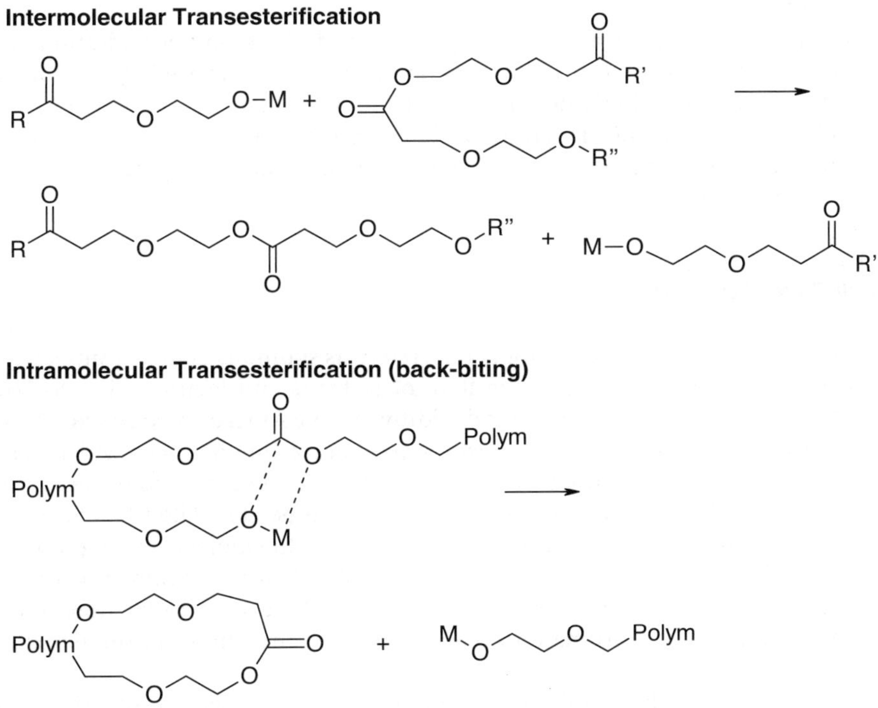

Scheme 5. Reaction schemes for intermolecular and intramolecular transesterification reactions

degradation of the polymer chain and the formation of cyclic oligomers [41]. Both types of transesterification reaction broaden the molecular weight distribution (MWD).

As displayed in the proposed scheme, each intramolecular transesterification randomly breaks the polymer chain. In this way, an attack on the polymer chain leads to a free residual polymer and a new randomized, modified polymer. Consequently, an original copolymer with a block-like structure would be converted to a randomized copolymer after undergoing n transesterifications [42, 43].

Parameters that influence the number of transesterifications are temperature, reaction time, and type and concentration of catalyst or initiator [44]. Depending on the metal used, the initiator is more or less active towards side-reactions such as transesterification reactions [44, 45]. The relative reactivity of different metal alkoxide initiators towards chains already formed has been reported to be: $Bu_2Sn(OR)_2 > Bu_3SnOR > Ti(OR)_4 > Zn(OR)_2 > Al(OR)_3$ [44].

The lactide configuration influences the extent of the transesterification reactions taking place during polymerization [46]. The contribution of transesterification processes in the case of D,L-lactide (D-LA) was found to be considerably higher than that observed when L,L-lactide (L-LA) was polymerized. The difference in the number of side-reactions was attributed partly to the polymer chain stiffness. The poly(D-LA) is more flexible than the poly(L-LA) due to the atactic lactide blocks.

When ε-CL and L-LA are block copolymerized, the monomer addition sequence is very important. AB block copolymers can be prepared by ROP with $SnOct_2$ as catalyst and ethanol as initiator provided that ε-CL is polymerized first [47]. If the L-LA block is synthesized first and the hydroxy-terminated macromer formed is used to initiate polymerization of ε-CL, the polymer formed is totally randomized.

2.1.2
Tin(II) 2-Ethylhexanoate

Tin(II) 2-ethylhexanoate, commonly referred to as stannous octoate [$Sn(Oct)_2$], is a frequently used catalyst in the ROP of lactones and lactides [42, 48–52]. $Sn(Oct)_2$ has been approved as a food additive by the American Food and Drug Administration (FDA). The mechanism of polymerization has been widely discussed. Despite several proposals [11, 47, 53, 54] over a long period of time, it is not until now that the ROP mechanism is about to be elucidated [55–58] The $SnOct_2$ is not thought to be the actual initiator since the molecular weight does not depend on the monomer-to-$SnOct_2$ molar ratio. The most promising mechanism is a coordination-insertion mechanism where a hydroxy functional group is thought to coordinate to $SnOct_2$, forming the initiating tin alkoxide complex.

Investigations of the coordination-insertion mechanism have resulted in two slightly different reaction pathways. Kricheldorf and coworkers have proposed a mechanism [11, 57] where the co-initiating alcohol functionality and the mon-

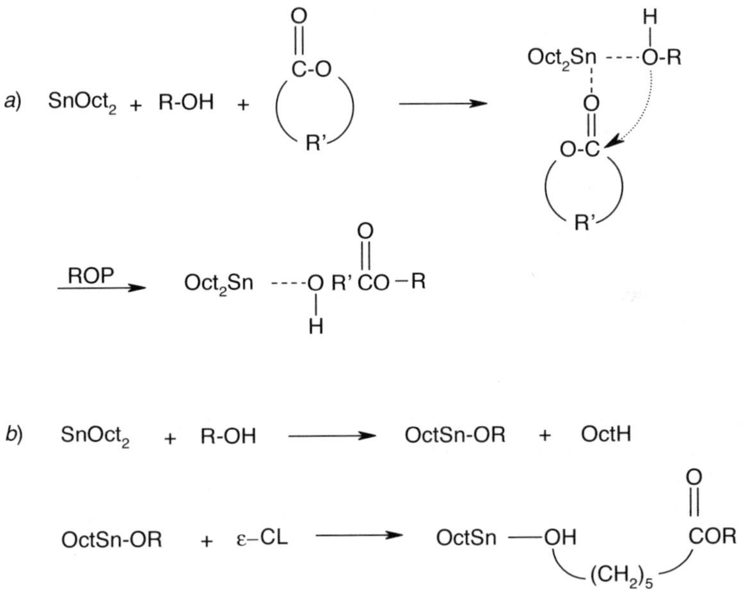

Scheme 6. The main ROP mechanism proposals with Sn(Oct)$_2$ as catalyst, *a*) complexation of a monomer and alcohol prior to ROP and *b*) formation of a tin-alkoxide before ROP of ε-CL

omer are both coordinated to the SnOct$_2$ complex during propagation. Penczek and coworkers have presented a mechanism [53] where the SnOct$_2$ complex is converted into a tin alkoxide before complexing and ring-opening of the monomer. Direct observation of this tin alkoxide complex has been reported by using MALDI-TOF spectroscopy for both lactide [56] and ε-CL [55] polymerization. Scheme 6 shows the two different proposals.

The SnOct$_2$ catalyst is a strong transesterification agent, and the resulting copolymers normally have a randomized microstructure [40]. An increase in reaction temperature or reaction time increases the amount of transesterification reactions.

The ROP of lactide with SnOct$_2$ is fairly slow and it is desirable for economic and commercial reasons to increase the rate of polymerization. The addition of an equimolar amount of triphenylphosphine increases the rate and, as an additional advantage, this compound delays the occurrence of the undesirable backbiting reactions [59].

2.1.3
Aluminum Tri-Isopropoxide

ROP initiated with aluminum tri-isopropoxide has been extensively investigated by several research groups [41, 44, 60–64] since it yields well-defined polymers through living polymerization [65, 66]. A living polymerization is a chain polymerization which proceeds in the absence of the kinetic steps of termination or chain transfer [65].

Polymerization with aluminum tri-isopropoxide is assumed to proceed through a coordination-insertion mechanism, which consists of monomer complexation to the active species and insertion by rearrangement of the covalent bonds. The mechanism leads to cleavage of the acyl-oxygen bond of the monomer and of the metal-oxygen bond of the propagating species. The propagation is characterized by the almost total absence of side-reactions such as transesterification reactions, at least until complete monomer conversion has occurred [67–69]. Some results indicate that transesterification reactions may take place during the polymerization of L-LA [69]. However, the main rearrangement of the polymer occurs when the monomers are completely consumed. The initiator is active at low temperatures (reaction temperatures of 0–25 °C are often reported) and the initiator is preferentially used in solution polymerization.

Most metal alkoxides are aggregated in solution and, as a result, an induction period during which the initiator is rearranged to form the active species often characterizes the polymerization. Only a few of the M-OR bonds are not involved in coordination to the metal atom and can consequently behave as initiation centers. The type and size of the aggregates depend on the solvent polarity, the nature of the substituents, and the presence of coordinative ligands such as amines [70] and alcohols [71, 72] The groups involved in coordinative aggregation are not active in propagation. Significant advances in the understanding of the "coordination-insertion" ROP mechanism have been made through the kinetic studies of Duda and Penczek [10, 41, 70, 71].

Recently, systems have been developed where the aluminum alkoxide is covalently bonded to solid porous silica [73]. This system takes advantage of the exchange reaction between the alkoxide and the hydroxy-terminated free molecules to produce a catalytic process, i.e., to produce a larger number of polymer chains than aluminum complexes present. The initiator/catalyst used can easily be recovered by filtration and recycled. In addition, the polymers obtained are free from metal residues.

2.1.4
Tin(IV) Alkoxides

Monotin alkoxides, tin dialkoxides and cyclic tin alkoxides have been utilized as initiators in the ROP of cyclic esters. The tin alkoxides are known to form cyclic species during synthesis and the dibutyltin alkoxides are known to exist as monomers and dimers [74]. The cyclic tin alkoxides were originally studied because of their resistance towards hydrolysis [75]. The tin alkoxides have been reported to be effective transesterification catalysts initiating polymerization at moderate temperatures [76].

The tributyl derivatives have been thoroughly studied since they are easily synthesized by nucleophilic substitution of commercial tributyltin chloride,

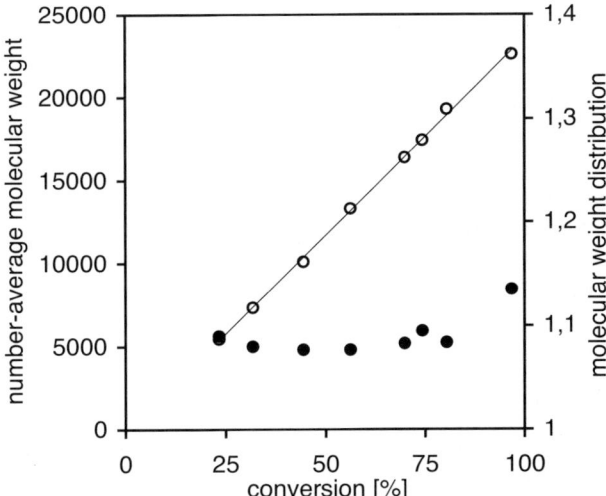

Fig. 2. The influence of L-LA monomer conversion on the number-average molecular weight (○) and the MWD (●). Polymerization conducted at 60 °C in chloroform with an initial monomer-to-initiator ratio of 100:1

they are easy to handle due to their moisture resistance, and they are relatively soluble in lactones [77].

The polymerization of lactones with tin alkoxides is thought to follow the coordination-insertion mechanism[77a]. The ring-opening of the monomer proceeds through acyl-oxygen cleavage with retention of the configuration. Tin(IV) complexes have been used to produce predominantly syndiotactic poly(β-hydroxybutyrate) [78, 79], macrocyclic poly(β-hydroxybutyrate) [80], poly(ε-CL), and polylactide [77, 76, 81].

The cyclic tin alkoxides have the additional advantage of offering a convenient synthetic pathway for the synthesis of macromers, triblock, and multiblock copolymers [81, 82]. Macromers from L-LA [83], ε-CL [84], and 1,5-dioxepan-2-one (DXO) [85] have been synthesized as well as triblock poly(L-LA-b-DXO-b-L-LA) [86] and multiblock copoly(ether-ester) from poly(THF) and ε-CL [87]. The polymerization proceeds by ring expansion and the cyclic structure is preserved until the polymerization is quenched by precipitation.

In a kinetic and mechanistic study on the polymerization of L-LA with a cyclic tin alkoxide [81] the number-average molecular weight increased linearly with increasing conversion and the MWD remained narrow (<1.15) throughout the polymerization reaction, Fig. 2.

The linearity of the plot of number-average molecular weight versus percentage conversion indicates that the amount of transfer reactions was low throughout the reaction. The increase in molecular weight was proportional to the degree of monomer conversion. The same characteristics have been observed for ROP of L-LA initiated by other cyclic tin alkoxides [83].

2.1.5
Tin(II) Alkoxides

Recently, tin(II) butoxide was used in the polymerization of L-LA [88]. The initiation is fast and quantitative and no transesterification or back-biting reactions are observed. The reaction proceeds with acyl-oxygen bond scission with retention of the configuration, and can be used both in bulk and solution (THF, 20–80 C) polymerization. It is possible to control the molecular weight in the range of 10^3 to 10^6 with a MWD of 1.15–1.85. The polymerization is very fast, $k_p = 5 \times 10^{-1}$ mol^{-1} L s^{-1}, with only the rare earth alkoxides being faster.

2.1.6
Lanthanide Alkoxides

ROP of lactones and lactides using lanthanide alkoxide-based initiators is a relatively recent discovery. The first example of lactone polymerization by lanthanide alkoxide complexes was reported in a DuPont patent written by McLain and Drysdale in 1991 [89]. In general, the activity of these catalysts is much higher than that determined for aluminum alkoxides, especially in lactide polymerization [90–92]. Polymers of relatively high molecular weight and narrow MWD are formed. The negative side-reactions such as macrocycle formation, transesterification, and racemization are absent.

Yttrium isopropoxide and yttrium 3-oxapentoxide initiators were the first lanthanide alkoxides described in the literature for the ROP of ε-CL [93]. The discovery of lanthanide-based initiator systems allowed the block copolymerization of ε-CL with compounds such as ethylene [94], tetrahydrofuran [95], L-LA [96], trimethylene carbonate [97], and methyl methacrylate [98]. This type of initiator has also been used to prepare poly(β-butyrolactone)s [99, 100].

3
Kinetics of Ring-Opening Polymerization

The kinetics of polymerization have been investigated in order to study the mechanism of controlled ROP. The results of the kinetic experiments have been utilized to understand the action of the initiator in more detail.

3.1
Kinetic Models

ROP reactions initiated with a metal alkoxide initiator are generally characterized by an equilibrium between the free and the aggregated metal alkoxide [10, 101, 102]:

$$\left(P_n^*\right)_m \xrightleftharpoons{K_{da}} m P_n^* \tag{1}$$

$$P_n^* + M \xrightarrow{k_p} P_{n+1}^* \qquad (2)$$

where P^*_n, (P^*_n) and M denote respectively non-aggregated active centers, aggregated active centers, and monomer, K_{da} is the aggregation equilibrium constant, k_p the propagation rate constant, and m the degree of aggregation. The aggregation causes a temporarily termination of the growing species, since the chains propagate only if they are non-aggregated. Due to the difference in reactivities between the aggregated compounds, the kinetics of polymerization are influenced.

In order to solve the kinetic equations corresponding to this system, Penczek and coworkers have recently proposed a method to determine the degree of aggregation from the curved plots of $\ln(k_{app})$ versus $\ln[I]_0$ [103]. The solution for the general case of the m-aggregate formation is:

$$(k_{app})^{1-m} = -m/K_{da}(k_p)^{m-1} + k_p[I]_0(k_{app})^{-m} \qquad (3)$$

and in the logarithmic form:

$$\ln(k_{app}) = 1/m \ln[I]_0 + C \qquad (4)$$

This equation allows a straight linear interpretation of the experimental data. When the logarithm of the apparent rate constant is plotted versus the logarithm of the initial initiator concentration, the slope of the line gives the external order in the initiator. The equation is valid for polymerization in which a fast reversible aggregation of the active centers takes place.

3.2
Kinetics of Lactide Polymerization

The kinetics of the L-LA polymerization have been investigated in chloroform at 60 °C [81] Fig. 3 shows the semi-logarithmic plot of $-\ln([M]/[M]_0)$ versus reaction time, t. $[M]_0$ is the initial lactide monomer concentration and $[M]$ the lactide concentration at a given reaction time t.

The linearity of the plot shows that the propagation was first order with respect to lactide monomer. The absence of an induction period indicates that the initiator was reactive from the beginning and that no rearrangement of initiator aggregates was necessary to form the active species. The nature of the metal, alkoxide groups, solvent, and temperature does not generally influence the first order in monomer [71, 104].

The kinetic equation describing this system is:

$$-d[M]/dt = k_{app}[M] \qquad (5)$$

Fig. 4 shows the apparent rate constant $(k_{app} = -\ln([M]/[M]_0)/t)$ as a function of the initial initiator concentration.

If the polymerization is concluded to be first order in initiator, the $k_{app}/[I]$ ratio must be constant as long as the number of active sites is independent of the

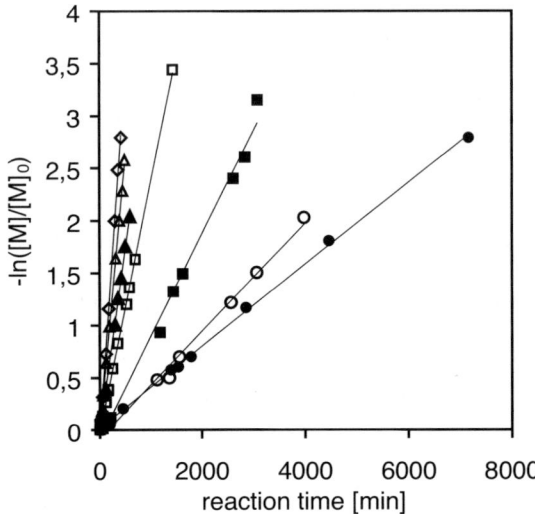

Fig. 3. The semi-logarithmic plot of $-\ln([M]/[M]_0)$ versus the reaction time for different monomer-to-initiator ratios. M/I=■ 25, △ 35, ▲ 50, ❏ 100, ■ 250, ○ 360, ● 400

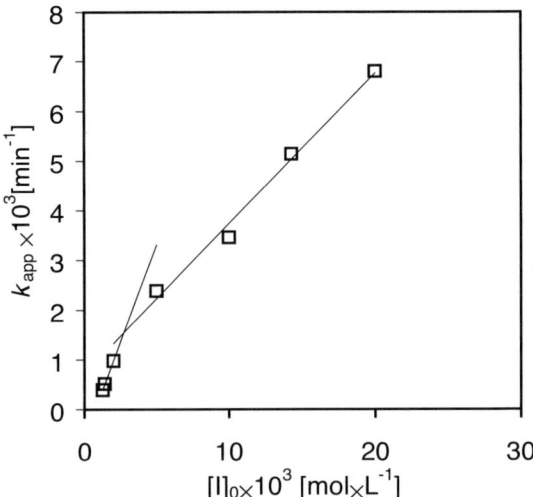

Fig. 4. Dependence of the apparent rate constant (k_{app}) on the initial initiator concentration $[I]_0$. Polymerization of lactide in chloroform at 60 °C

initiator concentration. When the polymerization proceeds with reversible aggregation, the plot is generally curved. Fig. 4 shows two distinct steps with different slopes, which reveal a change in the external order with respect to the initiator. It is thus clear that the tin alkoxide growing centers associate in chloroform. The change in the kinetic order in initiator has been attributed to the change in proportions of aggregated and non-aggregated species.

4
Macromolecular Architecture

Many recent advances in polymer synthesis have involved the development of new controlled polymerization systems proceeding via a variety of mechanisms. A number of architectures may be produced as a result of the great versatility of the ROP of cyclic esters. Different strategies have been applied for the design of new polymeric materials.

4.1
Homopolymers

Several factors are known to affect the ROP of cyclic esters. The main factors are the reaction conditions, i.e., the nature of the initiator, type of solvent and reaction temperature, and also the ring size of the monomer used and the substituents on the monomer ring [105–107] Cyclic esters of four-, seven-, and eight-membered rings polymerize, whereas the five- membered ester does not. In the case of six-membered rings, the polymerizability depends on the substituents [105].

4.2
Block Copolymers

Block copolymerization is one method of mixing chemically different polymers. The block copolymers have received much attention, since the different homopolymer properties are maintained in the block copolymer, and this allows easy modification of the polymer characteristics [63, 72, 86, 87, 108].

Block copolymers are most readily prepared by the sequential addition of monomers to systems polymerizing under living conditions. However, this approach is of limited importance since the monomers involved must all be capable of polymerizing by the selected propagating mechanism. The order of monomer addition must be such that the macroinitiator generated by the preceding monomer is capable of rapidly initiating the polymerization of the succeeding monomer. This condition limits the combination of monomers that can be used in block copolymerization. One example of this limitation is the copolymerization of DXO and L-LA initiated with aluminum isopropoxide [63]. The addition of DXO to the living system of poly(L-LA) macromers is not suitable, since the difference in reactivity ratio is too large [51], and the macromer formed when L-LA is polymerized cannot initiate polymerization of DXO. However, ε-CL macromers can initiate the polymerization of DXO and it is therefore possible to synthesize tri-block copolymers of DXO and ε-CL initiated with aluminum isopropoxide [64]. Fig. 5 shows a schematic presentation of a tri-block copolymer.

Fig. 5. A schematic presentation of an ABA tri-block copolymer with two A-blocks (gray circles with dark centers) and one B-block (gray circles with light centers)

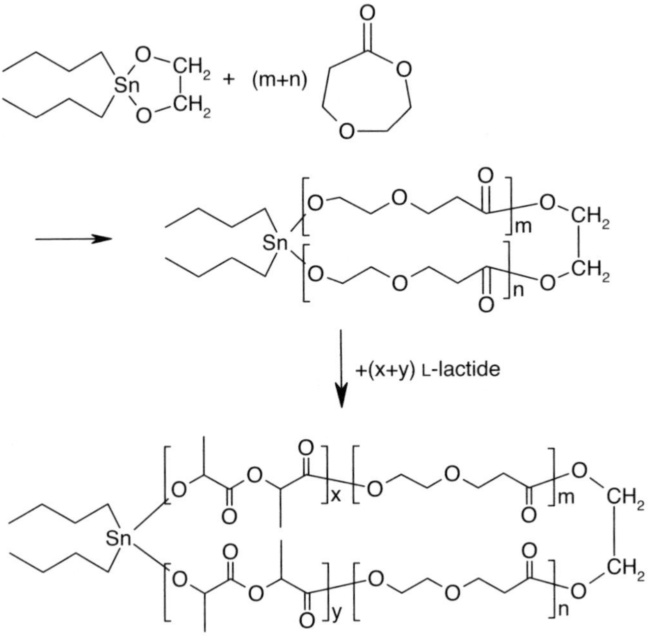

Scheme 7. The polymerization sequence for the synthesis of tri-block poly(L-LA-b-DXO-b-L-LA)

The large difference in reactivity ratio between L-LA and DXO makes it difficult to synthesize a tri-block copolymer, but the task can be carried out by using a cyclic tin alkoxide as initiator. The DXO macrocycle can initiate polymerization of L-LA and by ring expansion polymerization the two side blocks of L-LA are formed simultaneously. Scheme 7 shows the reaction pathway for the synthesis of tri-block copolymers from L-LA and DXO.

Thermoplastic tri-block copolymers are interesting since they possess novel properties different from those of the homo- or copolymers. The thermoplastic elastomers have many of the physical properties of rubbers, i.e., softness, resilience, and flexibility. The unique properties of this kind of copolymer are due to the microphase separation of the hard crystalline domains dispersed in a continuous amorphous matrix (Fig. 6). Such phase morphology provides a physical network of flexible chains cross-linked by crystalline microdomains. The advantages over natural vulcanized rubbers are that thermoplastic elastomers are readily soluble in an appropriate solvent and can be processed as thermoplastics [109].

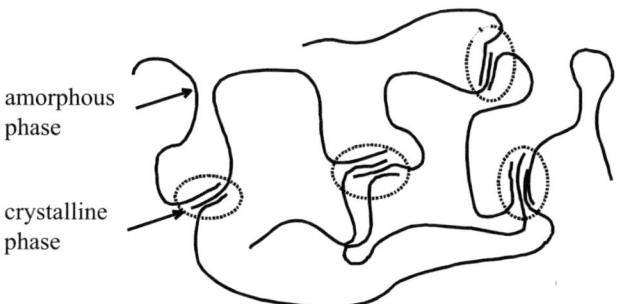

Fig. 6. Schematic picture of a block copolymer with one hard-phase-forming block and one soft-phase-forming block, giving rise to phase separation

One simple way of producing segmented (block-type) copolymers is reactive extrusion [110]. If two homopolymers are mixed and the mixture is subjected to high temperature, intermolecular transesterification reactions produce a block structure. If only a limited amount of transesterification reactions is allowed to take place, a block-type copolymer will be formed.

The preparation of prepolymers [111] or macromers with functional end groups, so called telechelic polymers, is another approach to structurally unconventional architecture. The functional end groups are introduced either by functional initiation or end-capping of living polymers, or by a combination of the two. In this way, monomers that are not able to copolymerize can be incorporated in a copolymer. Telechelic prepolymers can be linked together using chain extenders such as diisocyanates [112]. In this process, it is essential that the structure and end groups of the prepolymers can be quantitatively and qualitatively controlled [113].

4.3
Star-Shaped (Co)Polymers

Macromolecular structures such as star copolymers have been synthesized in the search for polymers with new mechanical and thermal properties and new degradation profiles. Fig. 7 shows a schematic representation of four-armed homo- and block copolymers.

Star-shaped polymers can be prepared by using a multifunctional initiator, e.g., pentaerythritol and a catalyst which initiates ROP of the selected monomer. A second approach is to use telechelic prepolymers that are linked together after polymerization.

Aliphatic star-shaped polyesters of L-LA have been synthesized [114, 115] with multifunctional hydroxy compounds as initiators. The crystallinity of the star-shaped poly(L-LA) was found to be higher than that of the corresponding linear counterpart. Star-shaped poly(L-LA) has also been block copolymerized with trimethylene carbonate/ε-CL [116] This resulted in a less brittle and considerably toughened material.

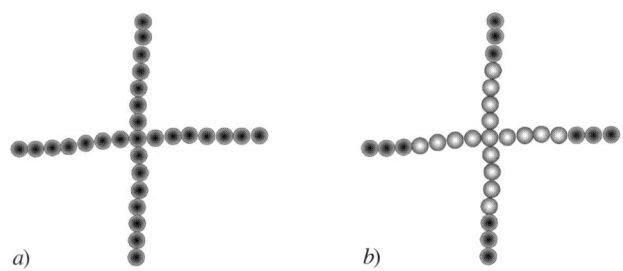

Fig. 7. Schematic representation of (*a*) a four-armed star-shaped homopolymer and (*b*) a star-shaped block copolymer

The unique thermal and solution properties of star-shaped block copolymers of poly(ethylene oxide)-poly(ε-CL) and poly(ethylene oxide)-poly(L-LA) have been studied [117]. The results, e.g., lower crystallinity and lower swelling with increasing degree of branching, may be used to influence the formulation, drug loading, and in vivo fate of drug and polymers in drug-delivery applications.

5
Biodegradable Polymers

Biodegradable polymers are receiving increasing attention for their use in a wide variety of surgical and pharmaceutical applications [118] as well as in disposable packages [119] Novel synthetic polymer materials may provide considerable improvement in medical applications due to their tailored thermal and mechanical properties and their decomposition to non-toxic products. Among various families of biodegradable polymers, aliphatic polyesters have a leading position since hydrolytic and/or enzymatic chain cleavage yields hydroxy carboxylic acids, which in most cases are ultimately metabolized [118, 120, 121]. Typical examples of synthetic, biodegradable polymers used in medical applications are polylactide [120], polyglycolide [122], and poly(ε-CL) [123]. The key properties, e.g., the rate of degradation, tensile properties, and surface chemical composition, can be optimized by copolymerization [124] or blending of homo- and/or copolymers.

5.1
Polymer Degradation

There are two principal ways by which polymer chains can be hydrolyzed, passively by chemical hydrolysis or actively by enzymatic reaction. The latter method is most important for naturally occurring polymers such as polysaccharides and poly(hydroxy alkanoate)s, e.g., polyhydroxybutyrate and polyhydroxyvalerate [121, 125]. Many synthetic aliphatic polyesters utilized in medical applications degrade mainly by pure hydrolysis [121].

There are several factors that influence the rate of degradation, including the type of chemical bond in the polymer back-bone, hydrophilicity, molecular weight, crystallinity, copolymer composition, and the presence of low molecular weight compounds [126, 127]. Other concerns are related to the loss of mechanical stability of polymers during degradation, which can be undesirable when it is too fast, or to the toxicity of high concentrations of degradation products. Many biodegradable polymers contain some kind of hydrolyzable bonds. Polymers containing anhydride [128] or ortho-ester [129] bonds are the most reactive ones with a fast rate of degradation. Ester bonds degrade somewhat more slowly, and carbonates [130] are almost totally resistant to hydrolysis.

5.1.1
Polyglycolide and Copolymers

Polyglycolide was one of the first synthetic polymers used as a degradable surgical suture [122]. Fig. 8 shows the glycolide monomer and polymer structures. This aliphatic polyester is biodegradable and exhibits negligible toxicity when implanted in tissue. It is also possible to fabricate a strong fiber of this polyester with satisfactory mechanical properties.

The polyester of glycolide has a melting point around 219 °C and is difficult to dissolve in most organic solvents. In order to produce materials suitable for specific medical applications, glycolide has been copolymerized with a number of monomers [131].

5.1.2
Polylactide and Copolymers

Due to the presence of two chiral centers, there are three forms of the lactide monomer (Fig. 9). Repeating units with different configurations have been used to produce stereocopolymers where the physical and mechanical properties and the rate of degradation are easily adjusted.

The most efficient way of preparing polylactides is ROP by coordination initiators [132]. This method usually allows a controlled synthesis leading to quite a narrow MWD. Polymerization of the different stereoforms results in materials with different properties. The polymers derived from the pure L-LA or D-LA

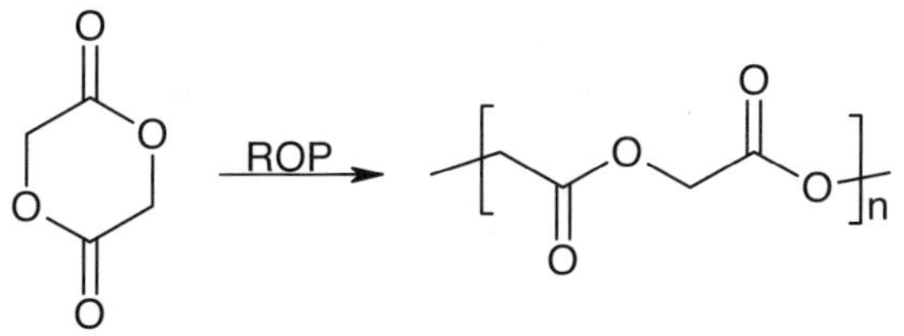

Fig. 8. The chemical structure of glycolide and the resulting repeating unit

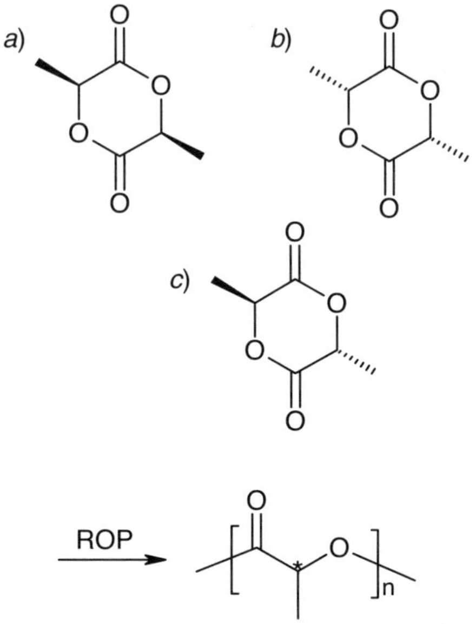

Fig. 9. Structure of the different stereoforms of the lactide monomer and the resulting repeating unit, the chiral center marked with *. *a*) L-LA, *b*) D,D-lactide, and *c*) *meso*-lactide

monomers are semi-crystalline, relatively hard materials with melting temperatures around 184 °C [133] and glass transition temperatures of about 55 °C [134]. The L,L-lactide and D,D-lactide are normally termed L-lactide and D-lactide, respectively. Polymerization of the *rac*-(D,L)-lactide and *meso*-lactide results in an amorphous material with a glass transition similar to that of the semi-crystalline counterparts [133]. Polylactides are highly sensitive to heat, especially temperatures higher than 190 °C. Heating these materials above this temperature results in a noticeable decrease in the weight-average molecular weight [133].

Poly(L-lactide) is biocompatible and when it is implanted in the body, it will in the course of time undergo hydrolytic scission to lactic acid, which is a natural intermediate in carbohydrate metabolism [120]. The in vitro degradation of poly(L-LA) is generally rather slow compared to the degradation of poly(D,L-LA). The higher degradability is probably due to the greater water absorption in the amorphous domains. Copolymers of L-lactide with several types of monomers, e.g., glycolide, or ε-CL, have been investigated [131]. Copolymerization is an important tool to modify the properties of the resulting copolymers and adjust them to the needs of a given application.

The crystallinity, brittleness, and melting point of poly(L-LA) can be decreased by incorporation of comonomer units such as 1,5-dioxepan-2-one (DXO). The large difference in reactivity ratio between the DXO and the lactides leads to a microstructure with a more block-like nature than is expected from a

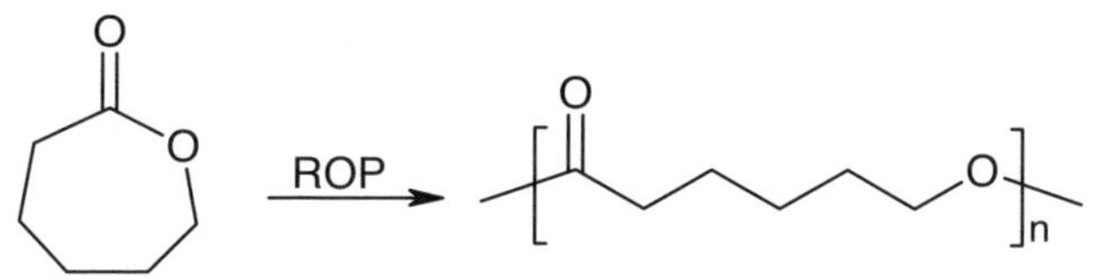

Fig. 10. The chemical structure of ε-CL and the resulting repeating unit

random copolymerization [51]. The copolymers have properties very different from those of the homopolymers. They show a low stiffness and high elasticity compared to poly(L-LA). The DXO/LA copolymers are interesting materials with possible applications in, e.g., the biomedical field [135]. The in vitro/in vivo degradation has been studied and it was found that the copolymer was hydrolyzed mainly by ester bond cleavage [136, 137].

5.1.3
Poly(ε-Caprolactone) and Copolymers

Poly(ε-CL) has been investigated thoroughly because of the possibility of blending this aliphatic polyester with a number of commercial polymers such as PVC and bisphenol A polycarbonate [17]. It is of interest as a packaging material and in biomedical applications since it is degradable and its degradation products are non-toxic. Fig. 10 shows the monomer structure and the resulting repeating unit.

The poly(ε-CL) material has a long degradation time, which is usually a disadvantage in medical applications. The in vivo degradation of poly(D-LA) was 2.8 times faster than that of the poly(ε-CL) chain cleavage under the same conditions [124]. Different approaches have been used to copolymerize ε-CL to increase the degradation rate. Copolymers of ε-CL and D-LA of all compositions degraded much more rapidly than their component homopolymers [124]. This observation has been attributed to morphological differences, specifically a reduction in crystallinity and a lowering of the glass transition temperature.

Random copolymers of ε-CL with 1,5-dioxepan-2-one (DXO) have been investigated [52, 138, 139]. The copolymers were crystalline up to a DXO content of 40%, and it was concluded that the DXO units were incorporated into the poly(ε-CL) crystals. The block copolymerization has also been investigated and the resulting material was shown to exhibit thermoplastic elastomeric properties [63].

6
Conclusions

A number of architectures may be produced as a result of the great versatility of the ROP of cyclic esters. Different strategies have been applied for the design of new polymeric materials. A careful selection of the appropriate initiator or catalyst for ROP of a specific system is crucial. The broad range of initiators and catalysts offer different advantages and possibilities. $Sn(Oct)_2$ is rather easy to use, but it is also a strong transesterification catalyst and it cannot therefore be recommended for the synthesis of advanced molecular structures.

For living ROP with the ability to control molecular architecture and weight, aluminum alkoxides can be used, the propagation being characterized by the almost total absence of side-reactions. The reaction is usually performed in solution at low temperatures. The sensitivity towards hydrolysis is however a limitation.

Tin alkoxides, on the other hand, are less sensitive to hydrolysis and can be used for controlled ROP and the synthesis of macromolecules with advanced architecture (tri-block, star, or comb polymers). Cyclic tin alkoxides offer a convenient pathway for tri-block copolymerization.

Lanthanide-based initiator systems offer a fourth possibility, permitting the block copolymerization of lactones with compounds such as ethylene,tetrahydrofuran, L-LA, trimethylene carbonate, and methyl methacrylate. Detrimental side reactions such as macrocyclic formation, transesterification, and racemization are absent and the reactions are extremely fast.

References

1. Mathisen T (1988) PhD thesis, KTH
2. Lundmark S (1989) PhD thesis, KTH
3. Löfgren A (1994) PhD thesis, KTH
4. Eklund M (1995) PhD thesis, KTH
5. Palmgren R (1997) PhD thesis, KTH
6. Carothers WH, Dorough GL, van Natta FJ (1932) J Am Chem Soc 54:761
7. van Natta FJ, Hill JW, Carothers WH (1934) J Am Chem Soc 56:455
8. Hill JW (1930) J Am Chem Soc 52:4110
9. Carothers WH, van Natta FJ (1930) J Am Chem Soc 52:314
10. Duda A, Penczek S (1994) Macromolecules 27:4867
11. Kricheldorf HR, Kreiser-Saunders I, Boettcher C (1995) Polymer 36:1253
12. Degée P, Dubois P, Jérôme R (1997) Macromol Chem Phys 198:1985
13. Pitt C, Gu Z-W (1987) J Controlled Release 4:283
14. Jérôme R, Henrioulle-Granville M, Boutevin B, Robin JJ (1991) Prog Polym Sci 16:837
15. Sosnowski S, Gadzinowski M, Slomkowski S (1996) Macromolecules 29:4556
16. Gadzinowski M, Sosnowski S, Slomkowski S (1996) Macromolecules 29:6404
17. Brode GL, Koleske JV (1972) J Macromol Sci – Chem A6(6):1109
18. Johns DB, Lenz RW, Luecke A, Lactones (1984) In: Ivin KJ, Saegusa T (eds), Ring-opening polymerization. Elsevier, London, 1:464
19. Löfgren A, Albertsson A-C, Dubois P, Jérôme R (1995) J Macromol Sci Rev Macromol Chem Phys C35(3):379
20. Mecerreyes D, Jérôme R, Dubois P (1999) Adv Polym Sci 147:1
21. Kricheldorf HR, Kreiser-Saunders I (1996) Macromol Symp 103:85
22. Saegusa T, Kobayashi S, Hayashi K (1978) Macromolecules 11:360
23. Cherdron H, Ohse H, Korte F (1962) Makromol Chem 56:179
24. Albertsson A-C, Palmgren R (1996) J Macromol Chem, Pure Appl Chem A33(6):747
25. Rozenberg BA (1992) Makromol Chem, Macromol Symp 60:177
26. Penczek S (2000) J Polym Sci, Polym Chem 38:1919

27. Kubisa P, Penczek S (1999) Prog Polym Sci 24:1409
28. Jérôme R, Teyssié P (1989) In: Eastmond GC, Ledwith A, Russo S, Sigwalt P (eds), Comprehensive polymer science. 1989, 3(1):501
29. Penczek S, Slomkowski S (1987) Progress in anionic ring-opening polymerization. In: Hogen-Esch T, Smid J (eds), Recent advances in anionic polymerization. Elsevier, New York, Chap 19:275
30. Jedlinski Z, Kurcok P, Kowalczuk M (1985) Macromolecules 18:2679
31. Hofman A, Slomkowski S, Penczek S (1984) Makromol Chem 185:91
32. Jedlinski Z, Walach W, Kurcok P, Adamus G (1991) Makromol Chem 192:2051
33. Sipos L, Zsuga M, Kelen T (1992) Polym Bull 27:495
34. Ito K, Hashizuka Y, Yamashita Y (1977) Macromolecules 10:821
35. Lundberg RD, Cox EF (1969) Lactones. In: Frish K, Reegen S (eds), Ring-Opening Polymerization. Marcel Dekker, New York, 2:247
36. Spassky N, Wisniewski M, Pluta C, Le Borgne A (1996) Macromol Chem Phys 197:2627
37. Ovitt TM, Coates GW (1999) J Am Chem Soc 121:4072
38. Kricheldorf HR, Bers M, Scharnagl N (1988) Macromolecules 21:286
39. Duda A, Florjanczyk Z, Hofman A, Slomkowski S, Penczek S (1990) Macromolecules 23:1640
40. Bero M, Czapla B, Dobrzynski P, Janeczek H, Kasperczyk (1999) J Macromol Chem Phys 200:911
41. Kowalski A, Duda A, Penczek S (1998) Macromolecules 31:2114
42. Grijpma DW, Pennings AJ (1991) Polym Bull 25:335
43. Gilding DK, Reed AM (1979) Polymer 20:1459
44. Dubois P, Ropson N, Jérôme R, Teyssie P (1996) Macromolecules 29:1965
45. Kricheldorf HR, Berl M, Scharnagl N (1988) Macromolecules 21:286
46. Bero M, Kasperczyk J (1996) Macromol Chem Phys 197:3251
47. In't Veld PJA, Velner EM, van de Witte P, Hamhuis J, Dijkstra PJ, Feijen J (1997) J Polym Sci A Polym Chem 35:219
48. Dahlmann J, Rafler G, Fechner K, Mehlis B (1990) Brit Polym J 23:235
49. Grijpma DW, Zondervan GJ, Pennings AJ (1991) Polym Bull 25:327
50. Kricheldorf HR, Meier-Haack J (1993) Macromol Chem 194:715
51. Löfgren A, Albertsson A-C (1995) J Macromol Sci Pure Appl Chem A32(1):41
52. Albertsson A-C, Gruvegård M (1995) Polymer 36:1009
53. Kowalski A, Duda A, Penczek S (1998) Macromol Rapid Commun 19:567
54. Schwach G, Coudane J, Engel R, Vert M (1997) J Polym Chem 35:3431
55. Kowalski A, Duda A, Penczek S (2000) Macromolecules 33:689
56. Kowalski A, Duda A, Penczek S (2000) Macromolecules 33:7359
57. Kricheldorf HR, Kreiser-Saunders I, Stricker A (2000) Macromolecules 33:702
58. Ryner M, Stridsberg K, Albertsson A-C, von Schenck H, Svensson M, (2001) Macromolecules 34:3877
59. Degée P, Dubois P, Jacobsen S, Fritz H-G, Jérôme R (1999) J Polym Sci Polym Chem 37:2413
60. Ouhadi T, Stevens C, Teyssié P (1975) Macromol Chem Suppl 1:191
61. Dubois P, Jacobs C, Jérôme R, Teyssié P (1991) Macromolecules 24:2266
62. Bero M, Kasperczyk J, Jedlinski Z (1990) Macromol Chem 191:2287
63. Löfgren A, Albertsson A-C, Dubois P, Jérôme R, Teyssié P (1994) Macromolecules 27:5556
64. Löfgren A, Renstad R, Albertsson A-C (1995) J Appl Polym Sci 55:1589
65. Quirk R, Lee B (1992) Polym Int 27:359
66. Gold L J (1958) Chem Phys 28:91
67. Dubois P, Jacobs C, Jérôme R, Teyssié P (1991) Macromolecules 24:2266
68. Duda A, Florjanczyk Z, Hofman A, Slomkowski S, Penczek S (1990) Macromolecules 23:1640

69. Montaudo G, Montaudo MS, Puglisi C, Samperi F, Spassky N, Le Borgne A, Wisniewski M (1996) Macromolecules 29:6461
70. Duda A, Penczek S (1991) Macromol Chem Macromol Symp 47:127
71. Duda A (1996) Macromolecules 29:1399
72. Jacobs C, Dubois P, Jérôme R, Teyssié P (1991) Macromolecules 24:3027
73. Miola C, Hamaide T, Spitz R (1997) Polymer 38:5667
74. Mehrothra RC, Gupta VD (1965) J Organometal Chem 4:145
75. Considine J (1966) Organotin chemistry VIII 5:263
76. Kricheldorf HR, Boettcher C, Tönnes K-U (1992) Polymer 33:2817
77. Kricheldorf HR, Sumbél M, Kreiser-Saunders I (1991) Macromolecules 24:1944
77a. von Schenck H, Reyner M, Albertsson A-C, Svensson M, submitted
78. Kemnitzer J, McCarty SP, Gross RA (1993) Macromolecules 26:6143
79. Kemnitzer J, McCarty SP, Gross RA (1993) Macromolecules 26:1221
80. Kricheldorf H, Lee S-R (1995) Macromolecules 28:6718
81. Stridsberg K, Ryner M, Albertsson A-C (2000) Macromolecules 33:2862
82. Kricheldorf HR, Langanke D (1999) Macromol Chem Phys 200:1174
83. Ryner M, Finne A, Albertsson A-C, (2001) Macromolecules in press
84. Kricheldorf HR, Eggerstedt S (1998) J Polym Sci Polym Chem 36:1373
85. Stridsberg K, Albertsson A-C (1999) J Polym Sci Polym Chem 37:3407
86. Stridsberg K, Albertsson A-C (2000) J Polym Sci Polym Chem 38:1774
87. Kricheldorf HR, Langanke D (1999) Macromol Chem Phys 200:1183
88. Kowalski A, Libiszowski J, Duda A, Penczek P (2000) Macromolecules 33:1964
89. McLain SJ, Drysdale NE (1991) US Patent 5 028 667
90. McLain S, Ford T, Drysdale N (1992) Polym Prep Am Chem Soc 33:463
91. Stevels W, Dijkstra P, Feijen J (1997) Trends Polym Sci 5:300
92. Stevels W, Ankoné M, Dijkstra P, Feijen J (1996) Macromolecules 29:6132
93. McLain S, Drysdale N (1992) Polym Prep J Am Chem Soc 33:174
94. Yasuda H, Furo M, Yamamoto H (1992) Macromolecules 25:5115
95. Nomure R, Endo T (1995) Macromolecules 28:5372
96. Stevels W, Ankoné M, Dijkstra PJ, Feijen J (1995) Macromol Chem Phys 196:1153
97. Shen Y, Shen Z, Zhang Y, Yao K (1996) Macromolecules 29:8289
98. Yasuda H, Ihara E (1995) Macromol Chem Phys 196:2417
99. Le Borgne A, Pluta Ch, Spassky N (1994) Macromol Rapid Commun 15:955
100. Xu J, McCarty SP, Gross RA (1996) Macromolecules 29:4565
101. Duda A, Penczek S (1994) Macromol Rapid Commun 15:559
102. Biela T, Duda A (1996) J Polym Sci Polym Chem 34:1807
103. Penczek S, Duda A (1991) Macromol Chem Macromol Symp 42/43:135
104. Ouhadi T, Hamitou A, Jérôme R, Teyssié P (1976) Macromolecules 9:927
105. Hall HK, Schneider AK (1958) J Am Chem Soc 80:6409
106. Saiyasombat W, Molloy R, Nicholson TM, Johnson AF, Ward IM, Poshachinda S (1989) Polymer 39:5581
107. Kurcok P, Matuszowicz A, Jedlinski A, Kricheldorf H, Dubois P, Jérôme R (1995) Macromol Rapid Commun 16:513
108. Albertsson A-C, Ljungqvist O (1986) J Macromol Sci – Chem A23(3):411
109. Noshay A, McGrath JE (1977) In: Block copolymers, overview and critical survey. Academic Press, New York, Chap 3:24
110. Stevels WM, Bernard A, van de Witte P, Dijkstra PJ, Feijen J (1996) J Appl Polym Sci 62:1295
111. Hiltunen K, Härkönen M, Seppälä J, Väänänen T (1996) Macromolecules 29:8677
112. Kylmä J, Seppälä J (1997) Macromolecules 30:2876
113. Hiltunen K, Seppälä J, Härkönen M (1997) J Appl Polym Sci 64:865
114. Kim SH, Kim YH (1994) Biodegradable Plastics and Polymers 464
115. Arvanitoyannis I, Nakayama A, Kawasaki N, Yamamoto N (1995) Polymer 36:2947
116. Grijpma D, Joziasse C, Pennings AJ (1993) Macromol Chem Rapid Commun 14:155

117. Choi YK, Bae YH, Kim SW (1998) Macromolecules 31:8766
118. Vainionpää S, Rokkanen P, Törmälä R (1989) Prog Polym Sci 14:679
119. Amass W, Amass A, Tighe B (1998) Polym Int 47:89
120. Kulkarni RK, Pani KC, Neuman C, Leonard F (1966) Arch Surg 93:839
121. Vert M, Li SM, Spenlehauer G, Guerin P (1992) J Mater Sci Mater Med 3:432
122. Frazza EJ, Schmitt EE (1971) J Biomed Mater Res Symp 1:43
123. Pitt CG, Chasalow FI, Hibionada YM, Klimas DM, Schindler A (1981) J Appl Polym Sci 26:3779
124. Pitt CG, Gratzl MM, Kimmel GL, Surles J, Schindler A (1981) Biomaterials 2:215
125. Holland S, Tighe B, Gould P (1986) J Controlled Release 4:155
126. Göpferich A (1996) Biomaterials 17:103
127. Mayer J, Kaplan D (1994) Trends Polym Sci 2:227
128. Albertsson A-C, Lundmark S (1990) Brit Polym J 23:205
129. Heller J, Sparer R, Zentner G (1990) Biodegrad Polym 45:121
130. Albertsson A-C, Eklund M (1995) J Appl Polym Sci 57:87
131. Grijpma D, Pennings AJ (1994) Macromol Chem Phys 195:1633
132. Leenslag JW, Pennings AJ (1987) Macromol Chem 188:1809
133. Jamshidi K, Hyon S-H, Ikada Y (1988) Polymer 29:2229
134. Kalb B, Pennings AJ (1980) Polymer 21:607
135. Edlund U, Albertsson A-C (1999) J Polym Sci Polym Chem 37:1877
136. Löfgren A, Albertsson A-C (1994) J Appl Polym Sci 52:1327
137. Löfgren A, Albertsson A-C, Zhang YZ, Bjursten L-M (1994) J Biomater Sci Polym Ed 6:411
138. Shalaby S, Kafrawy A (1989) J Polym Sci Polym Chem 27:4423
139. Shalaby S (1980) US Patent 4 190 720 (Ethicon Inc)

Received: January 2001

Degradable Polymer Microspheres for Controlled Drug Delivery

U. Edlund, A.-C. Albertsson

Department of Polymer Technology, Royal Institute of Technology, 10044 Stockholm, Sweden
e-mail: edlund@polymer.kth.se

Abstract. Controlled drug delivery technology is concerned with the systematic release of a pharmaceutical agent to maintain a therapeutic level of the drug in the body for a sustained period of time. This may be achieved by incorporating the therapeutic agent into a degradable polymer vehicle, releasing the agent continuously as the matrix erodes. This review is concerned with degradable polymers for use in controlled drug delivery with emphasis on the preparation, applications, biocompatibility, and stability of microspheres from hydrolytically degradable polymers.

Keywords. Controlled drug delivery, Drug release, Microspheres, Degradation, Erosion, Polylactide, Poly(glycolide-co-lactide), Poly(ε-caprolactone), Poly(hydroxyalkanoates) Polyanhydrides, Polycarbonates, Poly(orthoesters), Poly(1,5-dioxepan-2-one)

1	Introduction	68
1.1	Background	68
1.2	Polymer Degradation and Erosion	70
2	**Controlled Drug Delivery: State of Art**	72
2.1	The Concept of Controlled Drug Delivery	72
2.2	Release of Therapeutic Agents	74
2.3	Routes of Administration	75
2.4	Biocompatibility	76
3	**Polymers**	77
3.1	Aliphatic Polyesters	78
3.1.1	Polyglycolide, PGA	80
3.1.2	Polylactide, PLA	80
3.1.3	Poly(lactide-co-glycolide), PLGA	82
3.1.4	Poly(ε-caprolactone), PCL	84
3.1.5	Poly(3-hydroxybutyrate), PHB and other poly(hydroxyalkanoate)s	85
3.2	Polyanhydrides	88
3.3	Aliphatic Polycarbonates	91

| 3.4 | Poly(orthoesters), POE | 93 |
| 3.5 | Poly(1,5-dioxepan-2-one), PDXO | 96 |

4	**Microspheres**	98
4.1	General	98
4.2	Preparation	98
4.3	Controlled Release Applications	101

5	**Sterilization and Storage**	101
5.1	Sterilization	102
5.2	Storage	104

6	**Conclusions**	105

References 105

Abbreviations

PAA	poly(adipic anhydride)
PCL	poly(ε-caprolactone)
PDLLA	poly(D,L-lactide)
PDXO	poly(dioxepan-2-one)
PGA	poly(glycolide)
PHB	poly(hydroxybutyrate)
P(HB-co-HV)	poly(hydroxybutyrate-co-hydroxyvalerate)
PLGA	poly(lactide-co-glycolide)
PLLA	poly(L-lactide)
POE	poly(orthoester)
PTMC	poly(trimethylene carbonate)
RH	relative humidity
SEM	scanning electron microscopy
$Sn(Oct)_2$	stannous octoate, stannous-2-ethylhexanoate
Tg	glass transition temperature
Tm	melting temperature

1
Introduction

1.1
Background

The birth of synthetic polymers was the beginning of a new era in the field of human therapy. In drug devices polymers found use as excipients which adjusted the consistency of creams and liquid formulations, as plasma expanders, and as tablet coatings. Typically, cellulose derivatives, poly(vinylpyrrolidone), and poly(acrylates) were used. Generally, pharmaceutical research and development were focused on the synthesis and testing of new drug molecules, while less effort was spent on the final dosage form. Drugs were almost exclusively administered orally or injected, often at a site remote from the target tissue. With the development of more potent drugs, however, it became obvious that conventional therapeutic systems suffer from many drawbacks, including adverse effects, strongly fluctuating drug levels in the body, poor drug efficacy, and poor patient compliance. The concentration, duration, and bioavailability of the pharmaceutical agents could not be controlled. Controlled release technology was anticipated to circumvent these problems. The ambition was to maintain a therapeutic concentration of a drug in the body for a sustained period of time by releasing the agent in a predictable and controllable fashion. The first polymeric devices developed for controlled drug release date back to the early 1960s. Hydrogels, useful for drug delivery applications, were reported in 1960 [1]. Folkman and Long attracted attention when presenting a drug delivery system based upon the diffusion of small molecules through the wall of silicone rubber tubing [2].

At that point in time, the idea of using degradable polymers as resorbable matrices in medical applications was introduced. In the 1960s, poly(glycolic acid) and poly(lactic acid) was recognized as being highly interesting for use in temporary surgical implants and tissue repair [3–5]. Synthetic resorbable sutures from poly(glycolic acid) were first developed by American Cyanamid Co. in 1962 and they became commercially available under the name of Dexon™ in 1970 [6]. Vicryl™ sutures from poly(lactide-*co*-glycolide) (PLGA) emerged on the market 5 years later [7]. Since then, biodegradable polymers in general and aliphatic polyesters in particular have found use in a broad variety of medical applications: resorbable surgical sutures, prosthetics, artificial skin, dental implants, vascular grafts, pins, bone screws, as well as stents and plates for temporary internal fracture fixation [6–15]. Biodegradable polymers went from being merely of interest for research to playing a vital role in biomedical applications.

In the early 1970s Yolles et al. took an innovative step towards controlled drug delivery by achieving sustained release of cyclazocine dispensed in poly(lactic acid) sheets [16, 17]. Subsequent work demonstrated the systematic release of narcotic antagonists, fertility control agents, and anticancer drugs [17] and inspired other scientists to undertake research on the use of erodible polymer vehicles for controlled drug delivery [18–21]. Systems based on poly(lactic acid) were presented for long-term delivery of antimalarial drugs [19], contraceptives [20, 21], and eye drugs [22].

The belief that the therapeutic value of drugs could be improved by controlling the drug delivery led to the foundation of the ALZA Corporation. At this company, the desirability of developing drug delivery matrices eroding at the polymer-water interface was first formulated in the early 1970s. The first such system was reported in 1978 and was based on partial esters of vinyl acetate-maleic anhydride copolymers [23].

Since the field of controlled drug delivery emerged, continuously increasing numbers of scientists in academia and industry have adopted the challenge of designing polymeric systems for the controlled, systematic, or site-specific release of pharmaceutical agents. Many approaches have been adopted in attempts to obtain the optimal drug delivery device [24], including diffusion-controlled membranes (depot and monolithic systems), osmotic pumps, resorbable implants, hydrogels, ion-exchange materials, polymeric pro-drugs, and slowly dissolving matrices [24–27]. The use of biodegradable polymers has generally been favored over biostable polymers, since degradation of the matrices eliminates the need for surgical removal of the device after depletion [28].

Although research has been intense in the field of controlled drug delivery for the last 20 years, relatively few products are available on the market. In 1998, the worldwide market of drug delivery products was worth approximately US $30 billion, a modest share compared to the world market of pharmaceutics [29]. Commercialization of a biomedical device is a costly, lengthy process. Chemical and physical properties, biocompatibility, purity, and reproducibility of the product must be maintained through all steps of the scaled-up manufacturing process, packaging, and storage. Moreover, regulatory requirements of design, performance, and safety must be considered throughout the development phase of medical devices. The time span for the development of a new drug product from the time of discovery of a new therapeutic substance to its commercial introduction is now 10 years or more, compared to an average of 2 years in the 1950s. Even now, only a few biodegradable polymers have been approved by the US Food and Drug Administration (FDA) for use in a biological system [30]. Based on these limitations, only a limited number of drug delivery systems based on biodegradable polymers have so far been successfully commercialized. One example is Lupron Depot; one-month injectable microspheres of PLGA containing leuprorelin acetate (LH-RH agonist) for the treatment of endometriosis and prostatic cancer [31, 32].

1.2
Polymer Degradation and Erosion

Polymer degradation is defined as the chemical reactions resulting in a cleavage of main-chain bonds producing shorter oligomers, monomers, and/or other low molecular weight degradation products [33]. A polymer is considered to be biodegradable if the degradation is due to environmental action, either by biocatalytic processes (involving bacteria, fungi, enzymes, etc.) or by chemical and radical processes alone (hydrolysis, oxidation, UV irradiation) [34, 35]. Biodegradable polymers generate a continuously growing amount of interest and research in many fields, ranging from packaging, disposable items, and agriculture to surgical implants and drug delivery devices [34–37].

Biodegradable polymers used in the field of controlled drug delivery are typically degraded by hydrolysis [38]. In addition, some biomedical polymers are enzymatically degradable [35, 39]. Many factors are known to influence the biodegradation rate of a polymer [38, 39]. Polymer chemistry, molecular architecture, molecular weight, and morphology have a tremendous impact on the observed degradation rate [33, 38, 39]. The size, geometry, and porosity of the device are other important factors, as well as the surrounding conditions (e.g., pH and temperature) [40, 41].

Changes in the polymer properties accompany the degradation process. The most important parameter for monitoring degradation is the molecular weight. Loss of mechanical strength, crystallization, monomer formation, and pH changes are other events associated with degradation [33]. Hydrolytic degradation is caused by the reaction of water with labile bonds, typically ester bonds, in the polymer chain [38]. The reaction rate is intimately connected with the ability of the polymer to absorb water. Hydrophilic polymers take up large quantities of water and degrade faster than hydrophobic matrices [33].

Erosion is defined as the physical disintegration of a polymer matrix as a result of degradation [33, 42]. Upon incubation, water penetrates into the polymer matrix, advancing towards the center of the device, and induces chain scission. Once a sufficiently low molecular weight is reached, the degradation products formed will diffuse to and dissolve in the degradation medium and be transported away from the polymer matrix, i.e., erosion [33].

Depending on the erosion mechanism, the polymer is classified as either bulk eroding or surface eroding [33, 42]. If the water penetration proceeds faster than the matrix erodes, degradation will occur throughout the matrix and material will be lost from the entire polymer volume. This behavior is termed bulk erosion, but it is sometimes referred to as homogeneous erosion because mass loss proceeds at a more or less uniform rate throughout the matrix [43]. The size of the device remains constant even at later stages of degradation, but the microstructure within the bulk changes considerably. After erosion to a critical degree, the device eventually collapses [43]. The process of bulk erosion is schematically described in Fig. 1.

If water penetration is slow compared to the erosion process, mass loss is confined to the surface layers of the device only [42]. The size of the device gradually decreases but the bulk phase remains undegraded [33]. For ideal surface erosion, the erosion rate is directly proportional to the external surface area. Surface eroding devices are hence often preferred over bulk eroding materials for

Fig. 1. Bulk erosion of a degradable polymer device

Fig. 2. Surface erosion of a degradable polymer device

the sake of predictability. Surface erosion is, however, difficult to achieve. Many polymers are not sufficiently hydrophobic to keep water from penetrating and degrading the interior of the materials faster than the surface layer erodes [42]. Surface erosion is schematically illustrated in Fig. 2.

2
Controlled Drug Delivery: State of Art

2.1
The Concept of Controlled Drug Delivery

Traditional drug delivery systems, typically tablets or intravenous injections, administer the entire dose of drug in one portion resulting in high, sometimes close to toxic, plasma concentrations of drug, frequently leading to adverse reactions [25, 26]. Short duration times require repetitive administration, which is inconvenient for the patient, and lead to strongly fluctuating drug levels in the body. Moreover, conventional delivery is unpredictable and inefficient; often a high dose of drug is needed to ensure that the required amount of agent eventually reaches the site of action. The therapeutic effect is, however, achieved not simply by administration of the drug but rather by attaining a long-term appropriate drug concentration. The prolonged drug delivery system is therefore characterized by releasing the drug in a controlled fashion to maintain an appropriate, therapeutic plasma concentration for a long period of time, and by pro-

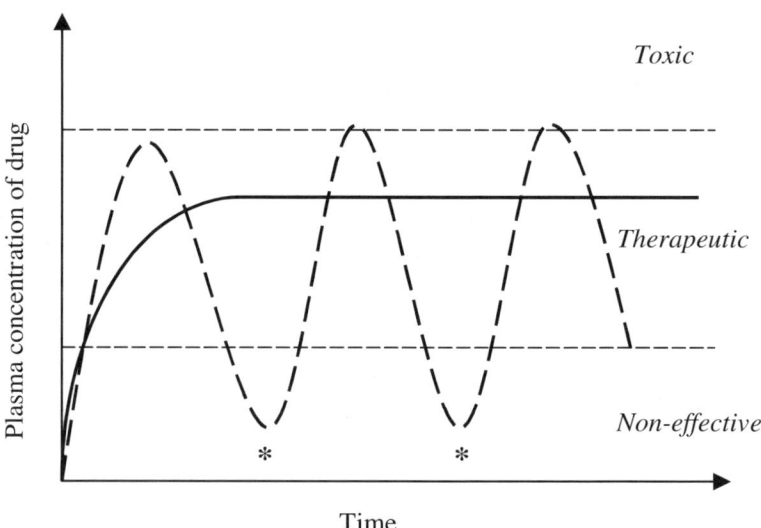

Fig. 3. The plasma concentration of drug in the patient as a function of time after administration: (- - -) traditional delivery system with repetitive administration (*), (–) prolonged delivery system

viding the drug only where and when it is needed. The traditional and prolonged systems are schematically compared in Fig. 3 [26].

Controlled drug delivery systems seek to mimic the prolonged system by providing therapeutic action for an extended period of time, by reducing adverse effects, and by optimizing patient compliance. In some treatments, however, pulsatile delivery may be desirable [44, 45]. Several approaches have been presented in the development of drug delivery devices [24–26, 46]. Among these systems, the administration route and the drug-release mechanism vary over a broad range [47]:

- Diffusion-controlled membranes exist in two categories; depot systems, in which the drug is totally encapsulated within a reservoir, and monolithic systems, where the drug is dispersed in a rate-controlling polymer matrix [25]. One commercially successful depot device is the Alza Ocusert for ocular delivery of pilocarpine in the treatment of glaucoma [25].
- Ointments, creams, and plasters are used for transdermal drug delivery. In the latter case, sustained release is accomplished by diffusion from a reservoir through a microporous membrane and into the skin [26, 48, 49]. Iontophoresis and electroporation has been reported to successfully promote transdermal delivery rates [50–52].
- Pumps are intended for implantation. An osmotic pressure differential triggers drug release from a reservoir embraced by a semipermeable membrane [26].
- Pro-drugs are systems comprising drug molecules chemically bonded to a polymer carrier. The drug converts to the active compound at a specific tissue site, often through the agency of an enzyme or receptor [53–56].
- Hydrogels are cross-linked hydrophilic polymers, highly swollen but not soluble in the surrounding medium [27]. Drug is released by diffusion from the network.
- Liposome systems [46, 57] are vesicles bounded by bimolecular phospholipides layers. Unilamellar systems contain a single layer where a drug is entrapped within the central compartment whereas multilamellar systems consist of several concentric layers.
- Polymeric carriers are biodegradable or water-soluble polymer matrices, typically in the form of colloidal-sized particles (microspheres or nanospheres), rods, or films. The active agent is entrapped within but not chemically bonded to the matrix. The drug is released in a sustained fashion as the polymer is dissolved or degraded, eroded, and finally resorbed [24, 30, 58–62].

Controlled release technology offers many advantages and can improve many treatments [24, 26, 30]. Drug delivery formulations may lead to an increased efficiency of labile drugs. Once released from the dosage form, the drug must pass through several physiological barriers before reaching the site of action and must then survive metabolic and chemical attacks. Labile drugs, typically peptides, proteins, and enzymes, may lose activity due to the local tissue environment, e.g., the acidity in the stomach [63]. A biodegradable polymer may thus act as a transient mask, protecting the therapeutic agent from physiological degradation until the desirable tissue is reached. Sustained release also improves patient compliance. The reduced need for repetitive dosing is a relief to many patients, including the elderly and those chronically ill. By placing the polymer implant adjacent to the surgical site, the drug delivery is enhanced in the appropriate tissues, and the efficacy of the therapeutic agent is improved. Potentially toxic drugs, typically anticancer agents, can be delivered in high concentrations to the tumor site, increasing the efficiency and at the same time minimizing systemic exposure and subsequent side-effects [55, 64]. A considerable potential for controlled delivery devices exists in public health programs in underdeveloped countries and there has been much interest in developing carriers for the delivery of vaccines and adjuvants [45, 65–67]. Sustained delivery of immunogens might result in an enhanced immune response eliminating the need for vaccination boosters. Alternatively, single-shot vaccines may be obtained from pulsatile delivery systems that release a burst of antigens [45]. Controlled delivery of insulin presents another important challenge. Diabetes is a growing problem in the industrial countries and the treatment is demanding to the patient, requiring daily subcutaneous injections. Oral administration of free insulin is not an option because of poor bioavailability and instability towards the enzymes in the gastrointestinal tract. Degradable polymers could serve to protect the insulin and release it in a prolonged and controlled fashion [63]. Controlled delivery has applications not only in human therapy but also in veterinary and agrochemical fields. Active substances from contraceptives to pesticides can potentially be released in a systematic fashion from a bioerodible polymeric carrier. This review will henceforth focus on drug delivery technology based on degradable polymeric carriers, and microspheres in particular.

2.2
Release of Therapeutic Agents

Several mechanisms may be responsible for the overall release of a therapeutic agent dispersed in a degradable polymer matrix [24, 60, 68]:
- Erosion of the polymer;
- Diffusion of the drug particles through the matrix;
- Dissolution of the drug in the surrounding medium.

Drug molecules entrapped within a degradable polymer matrix will be liberated and released as degraded material is lost from the matrix by erosion. Concomitantly, the concentration gradient is a driving force for the diffusion of drug molecules from the matrix to the surrounding medium. The third mechanism is most significant in the initial stages of incubation when drug molecules deposited on or near the matrix surface are lost by dissolution in the surrounding medium. The relative importance of these mechanisms for the overall release rate varies considerably from one system to another, depending on the polymer hydrophilicity, flexibility, degradation rate, molecular weight, crystallinity, and the matrix size, shape, and porosity [47]. For the vast majority of drug delivery systems, all mechanisms contribute to the overall release of drug, albeit with varying proportions [68]. The kinetics of drug release are also influenced by the physical properties of the drug, particularly its molecular weight and solubility in water. Diffusivity of a drug through a polymer barrier is dependent upon the solubility of the drug in the polymer, the size of the drug molecule, and its distribution throughout the matrix. Environmental conditions (e.g., pH and temperature of the surrounding medium) must also be considered [41, 69].

An early kinetic model was proposed by Higuchi for pure diffusion-mediated release of drug from a monolithic device, derived from Fick's second law of diffusion [70]. This relationship states that the amount of drug release is proportional to the square root of time. Peppas et al. introduced a generalized expression stating that the fractional release is proportional to t^n, where the exponent is indicative of the transport mechanism [71]. Over the years, extended and modified models and expressions have been presented [47]. Kinetic expressions for the release from biodegradable matrices have been derived, taking into account the changes in molecular weight and diffusion coefficient over time [72, 73]. According to Higuchi's theory, the drug release rate will diminish with time because the concentration gradient is continuously decreasing [70]. This is in many cases, however, not considered as being the optimal drug release behavior. Instead, "zero-order" release is desirable, where a constant release rate is maintained throughout the process [74]. True zero-order release is almost unattainable, although many attempts have been made to mimic this behavior [74–76]. It has been postulated that a truly surface-eroding polymer matrix may provide zero-order drug release [77].

2.3
Routes of Administration

A therapeutic substance can be administered to a patient via a number of pathways. Oral delivery (by the mouth) was among the very first drug delivery methods used in the pharmaceutical industry [26]. Drugs may also be transferred to the alimentary channel via sublingual (under the tongue) or rectal (to the rectum) administration.

Any way of administration other than through the digestive system is referred to as parenteral administration. Important parenteral pathways are listed in Table 1 [78]. The local tissue environment differs throughout the body, e.g., in terms of pH and enzyme activity. The site of administration may thus influence the performance of a drug delivery device.

Table 1. Parenteral routes of drug administration

Sites of parenteral administration	
Intraocular	in the eye
Intravenous	injection in a vein
Intra-arterial	injection in an artery
Intraspinal	injection in the spine
Intraosseous	in a bone

Table 1. (continued)

Intra-articular	in a joint/joint fluid
Topical	on the skin surface
Subcutaneous	between skin layers
Transdermal	through the skin
Nasal	to the nose
Aural	to the ear
Intrarespitory	to the lung
Vaginal	to the vagina

Table 2. General requirements of a biocompatible polymer

Prerequisites of biocompatibility
Non-toxic
Non-cancerogenic
Non-mutagenic
Non-allergenic
Free from contaminants (e.g., additives, solvents, and synthesis residues)
Biocompatible degradation products
Bioresorbability
No adverse immunological responses

2.4
Biocompatibility

The response reaction of the host to a foreign material remaining in the body for an extended period of time is a concern. Thus, any polymeric material to be integrated into such a delicate system as the human body must be biocompatible. Biocompatibility is defined as "the ability of a material to perform with an appropriate host response in a specific application" [79]. The concept include all aspects of the interfacial reaction between a material and body tissues: initial events at the interface, material changes over time, and the fate of its degradation products. To be considered biocompatible, a biodegradable polymer must meet a number of requirements, given in Table 2.

The term "bioresorbable" refers to polymers which degrade into products that can be eliminated from the body through natural pathways or, even better, which are involved normally in metabolic pathways [13]. Toxicity does not necessarily stem from the polymer itself or its fragments, but may arise from the presence of synthesis residues such as solvents, catalysts, monomers, and stabilizers [80].

Various standards and procedures exist for the evaluation of the biological and immunotoxicity response of an implant [81] from the point of view of biocompatibility. Acute toxicity screening and in vivo implantation tests are fundamental in this respect. Cytotoxicity testing to detect the biological activity of the material on a mammalian cell monolayer is often the first step in assessing biocompatibility of a device. An international standard on the biological evaluation of medical devices, ISO 10993, was issued in 1992 by the International Standards Organization (the ISO). FDA adopted a modified version in 1995. These standards serve as guidelines for adequate evaluation of biocompatibility [81].

The sensitivity to irritation is different for different tissues in the body. Biocompatibility is therefore highly related to the injection site. For instance, the rabbit eye is a highly sensitive animal model for biocompatibility studies [82, 83]. Surface topography is another important parameter of biocompatibility [79]. Sharp edges or corners may cause irritation and enhance the local tissue response [40].

3
Polymers

Based on their behavior in living tissue, polymeric biomaterials can be divided into two groups; biostable and biodegradable. Biostable polymers are used when permanent aids are needed, e.g., as prostheses [13]. Biostable polymers, typically polyethylene and poly(methyl methacrylate), should be physiologically inert in tissue conditions and maintain their mechanical properties for decades [11].

Biodegradable polymers are intended for temporary aids, such as sutures, tissue-supporting scaffolds, and drug delivery devices [13]. Polymers within this group retain their properties for a limited period of time and then gradually degrade into soluble molecules that can be excreted from the body [11]. As previously mentioned, biocompatibility of the polymer and its degradation products is essential for any material which will be in contact with living tissue [79].

Biodegradable polymers are preferred for drug delivery applications, since the need for surgical removal of the depleted device is eliminated. Although the number of biodegradable polymers is large, only a limited number of polymers is suitable for drug delivery applications. Suitable candidates must not only be biodegradable but also fit the high prerequisites of biocompatibility. In addition, a polymer should ideally offer processability, sterilizability, and storage stability if it is to be useful for biomedical applications [15].

Polymers mainly investigated for drug delivery applications are of either natural or synthetic origin. The former group includes:
- Polysaccharides, e.g., dextran or cellulose [84, 85];
- Chitin [86];
- Chitosan [86, 87];
- Proteins (e.g., collagen, fibrin, gelatin, albumin) [88, 89].

Synthetic degradable polymers investigated for controlled drug delivery applications include:
- Aliphatic polyesters;
- Poly(glycolide), PGA [6, 15, 61, 90];
- Poly(lactide), PLA [15, 17, 20, 21, 61, 90–95];
- Poly(glycolide-co-lactide), PLGA [94–104];
- Poly(ε-caprolactone), PCL [20, 61, 105, 106];

- Poly(3-hydroxybutyrate), PHB [107–109];
- Poly(3-hydroxybutyrate-co-3-hydroxyvalerate), P(HB-co-HV); [107, 109, 110]
- Polyanhydrides [75, 111–119];
- Aliphatic polycarbonates [120, 121];
- Poly(orthoesters) [122–128];
- Poly(1,5-dioxepan-2-one), PDXO [129–131];
- Poly(amino acids) [132, 133];
- Poly(ethylene oxide) [134–137];
- Polyphosphazenes [138–140].

Many efforts have been made to obtain new polymer systems having the desired mechanical and physicochemical properties for a specific medical application. Methods of tailoring polymers include molecular architecture, copolymerization, and homopolymer blending [13, 141].

3.1
Aliphatic Polyesters

Aliphatic polyesters have been known for a long time and were investigated as part of Carother's pioneering studies on polymerization in the 1930s [142]. The family of aliphatic polyesters has been by far the dominating choice for materials in degradable drug delivery systems. Homo- and copolymers derived from glycolic acid or glycolide, lactic acid or lactide, ε-caprolactone, and 3-hydroxybutyrate have been given special attention. Structures of these polymers are presented in Fig. 4.

Aliphatic polyesters degrade chemically by hydrolytic cleavage of the backbone ester bonds [38, 92, 93, 143–145] and by enzymatic promotion [35, 146]. Hydrolysis can be catalyzed by either acids or bases [38]. Polyester hydrolysis is schematically illustrated and exemplified for PLA in Fig. 5. Carboxylic end groups are formed during chain scission, and this may enhance the rate of further hydrolysis. This mechanism is denoted "autocatalysis" [147] and makes polyester matrices truly bulk eroding [38, 43]. Degradation products are resorbed by the body with a minimal reaction of the tissues [8, 15, 95, 148].

Polyesters can be synthesized by step-wise polycondensation of hydroxy acids or of diols and diacids, or by ring-opening polymerization (ROP) of the cyclic lactone [13, 38, 92]. Carother's pioneering studies in the 1930s involved polyester synthesis by polycondensation [142]. Today, ring-opening polymerization is the preferred method of producing high-molecular weight polyesters, because milder conditions and shorter reaction times can be used [149, 150]. To obtain high conversion, polycondensation requires long reaction times and high temperatures. Furthermore, there is a strict need for removal of reaction by-products and the use of a stochiometric balance of monomers, since the esterification reaction is reversible in nature. This can be avoided by employing ring-opening polymerization. The ring-opening polymerization of lactones, investigations of mechanisms and initiators are extensively documented in the literature [92, 144,

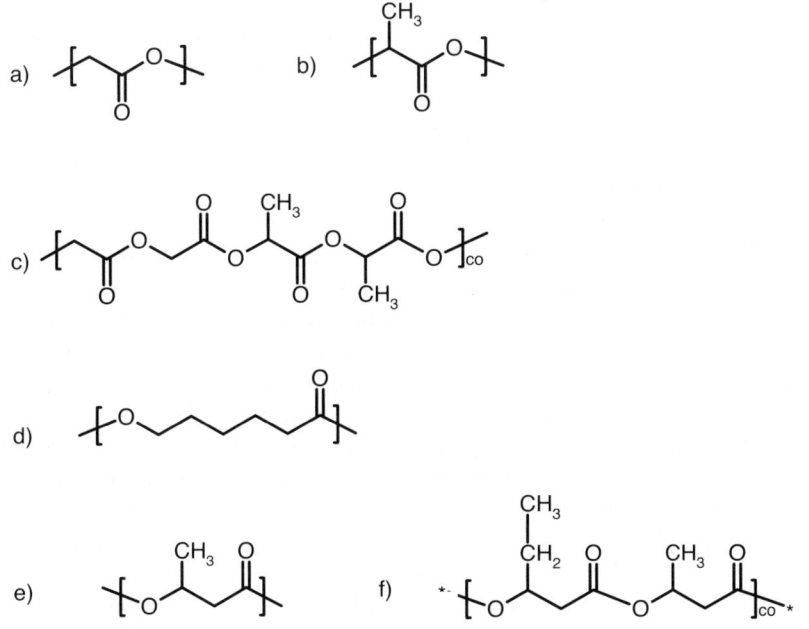

Fig. 4. The molecular structures of some aliphatic polyesters: *a*) poly(glycolide), PGA; *b*) poly(lactide), PLA; *c*) poly(lactide-*co*-glycolide), PLGA; *d*) poly(ε-caprolactone), PCL; *e*) poly(hydroxybutyrate), PHB; and *f*) poly(hydroxybutyrate-*co*-hydroxyvalerate), P(HB-*co*-HV)

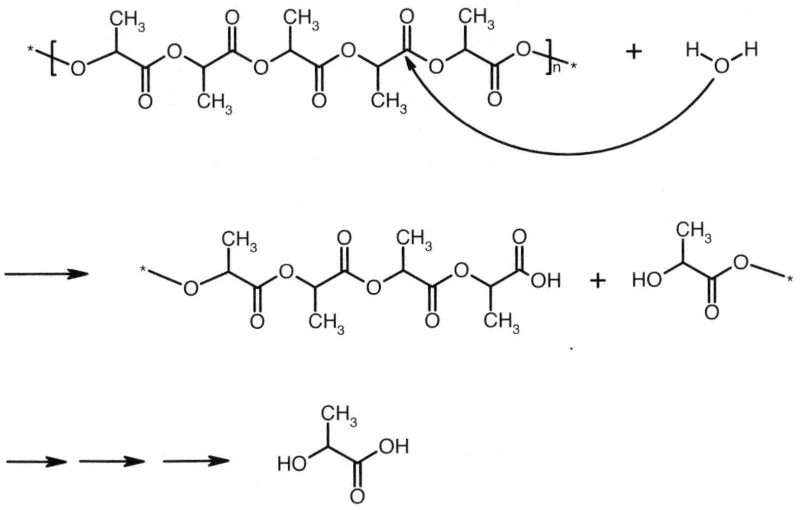

Fig. 5. Schematic hydrolytic degradation of PLA, an aliphatic polyester

149–159]. Synthetic routes include anionic, cationic, zwitterionic, and coordination polymerization. A wide range of organometallic compounds has been proven as effective initiators/catalysts for ROP of lactones; Lewis acids (e.g., $AlCl_3$, BF_3, and $ZnCl_2$) [150], alkali metal compounds [160], organozinc compounds [161], tin compounds of which stannous octoate [also referred to as stannous-2-ethylhexanoate or tin(II) octoate] is the most well known [162–164], organoacid rare earth compounds such as lanthanide complexes [165–168], and aluminum alkoxides [169]. Stannous-2-ethylhexanoate is one of the most extensively used initiators for the coordination polymerization of biomaterials, thanks to the ease of polymerization and because it has been approved by the FDA [170].

Enzyme-catalyzed polyester synthesis has attracted attention in the recent years. Enzymatic polymerization is attractive from the point of view of molecular architecture. Biocompatible, pure, and well-defined polymer structures may be obtained as the method is highly selective and proceeds without side-reactions under mild conditions. Enzymatic catalysis of polyester polymerization of both the condensation and the ring-opening type has been reported. Lipases from the family of *Pseudomonas* were found to be active for the polymerization of various lactones, including ε-caprolactone (ε-CL), γ-CL, γ-valerolactone, β-butyrolactone, δ-valerolactone, 8-octanolide, 11-undecanolide, and 12-dodecanolide [171–173]. High molecular weight polylactide has been obtained by ROP catalyzed by lipase PS at a bulk temperature of 80–130 °C [174]. The mechanism of lipase-catalyzed polymerization of lactone has been suggested to involve both ring-opening polymerization and linear condensation.

3.1.1
Polyglycolide, PGA

PGA is commonly obtained by ring-opening polymerization of the cyclic diester of glycolic acid, glycolide [6,8]. PGA is a hard, tough, crystalline polymer with a melting temperature of ≈225 °C and a glass transition temperature, Tg, of 36 °C [6]. Unlike closely related polyesters such as PLA, PGA is insoluble in most common polymer solvents [6]. PGA has excellent fiber-forming properties and was commercially introduced in 1970 as the first synthetic absorbable suture under the trade name Dexon™ [6]. The low solubility and high melting point of PGA limits its use for drug delivery applications, since it cannot be made into films, rods, capsules, or microspheres using solvent or melt techniques.

3.1.2
Polylactide, PLA

Early reports of the biomedical use of poly(lactic acid) date back to the 1960s [3, 175]. Since then, PLA has gained widespread application in the medical field, for use in sutures [8], drug delivery devices [15, 17, 24, 59–90], prosthetics, scaffolds, vascular grafts, and bone screws, pins and plates for temporary internal fracture fixation [10–13]. Good mechanical properties and the fact that it degrades into non-toxic products explain the popularity of PLA [9–11, 38, 61, 145]. In addition, PLA has been approved by the FDA for medical use and is commercially available in a variety of grades [30]. The biocompatibility of PLA is well documented. A good number of studies agree that PLA is completely resorbable, triggering no or very mild and transient adverse tissue responses [9, 10, 95, 148, 170]. Hydrolytic degradation eventually generates the monomer lactic acid, which is metabolized via the tricarboxylic acid cycle and subsequently eliminated as CO_2 via the respiratory system [3, 9]. The hydrolytic scission of PLA proceeds at a slower rate than that of the more hydrophilic PGA. The role of enzymes in the in vivo degradation of PLA has been intensely debated. It is clearly shown that PLA can be degraded by the action of various enzymes, such as proteinase K [146]. Many studies, however, indicate that there are no significant differences between in vivo and in vitro degradation, excluding the effect of enzymes [12, 143, 176]. The differences between the in vivo and in vitro degradation rates claimed by some authors may instead be explained by an increase in polymer chain mobility due to lipid components present in living tissue [41].

PLA is commonly synthesized by the ROP of lactide, the diester of lactic acid [92, 149]. The lactide has two asymmetric carbons and thus exists as the optically active L- and D-forms or as the racemic D,L-form. The pure enantiomeric poly(L-lactide) (PLLA) is naturally occurring and is semicrystalline because of its stereoregular structure. PLLA is a relatively hard material with a crystalline melting temperature in the range of 170–180 °C and a Tg of ~60–67 °C [15]. Polymerization of the diastereoisomer (D,L-lactide) or a racemic mixture of D,D-lactide/L,L-lactide yields an amorphous poly(D,L-lactide) (PDLLA) [13, 92]. PDLLA has a Tg in the region of 50–60 °C. Since polymers from lactic acids have glass transition temperatures above body temperature, these matrices are stiff with little elasticity in the body and are somewhat brittle at room temperature [177]. Polymer solubility in common organic solvents is important for the fabrication of drug delivery systems. PLA is soluble in halogenated hydrocarbons, ethyl acetate, tetrahydrofuran, dioxane, and a few other solvents [15].

The commercial interest in PLA is continuously growing. This is governed by the recent advances in processing, and engineering of the product properties, but most of all by the recent developments in manufacturing the monomer from renewable resources [178]. Up until the mid 1990s, PLA was produced from a petrochemical feedstock and the high product price limited its applicability. Now, the monomer can be economically obtained by bacterial fermentation of D-glucose from corn and the lowered market price of PLA opens up a range of new applications, including food packaging and other disposable items.

Copolymerization and blending of PLA has been extensively investigated as a useful route to vary the chemical structure of the copolymer over a wide range to obtain a product with a particular combination of desirable properties. Copolymerization of L-LA with related cyclic esters such as glycolide, D-LA, ε-CL, ε-decalactone, or with TMC yields degradable polymers that are highly interesting for drug delivery applications [20, 90, 144, 179–182]. Various copolymers of PLA and poly(ethylene glycol), PEG, have been reported. The introduction of PEG segments enhances both the hydrophilicity and rate of degradation of the material. Block copolymers of PEG and PLA have been appreciated for their ability to form micelles. This core-shell structure has a great potential as drug-loaded systems [136, 183–186]. PLA has been shown to form miscible blends with P(HB-co-HV), PHB, PCL, poly(mandelic acid), and poly(propylene fumarate) in melt and solution [141, 187].

PLA has been investigated for the systematic delivery of a broad variety of therapeutic agents. Early reports include the use of PLA for the delivery of contraceptive steroids [21], narcotic antagonists [17], and antimalarial agents [19]. Zero-order release of L-methadone (a narcotic antagonist) was obtained when using a mixed matrix composed of PLA, P(LA-co-CL), and PLGA [188]. Apparent zero-order release was also obtained from microspheres made from oligomers of PDLLA containing aclarubicin hydrochloride (an anti-cancer agent) [72]. Other anti-tumor systems studied include 5-fluoro-2'-deoxyuridine pro-drugs in PLLA microspheres [189] and taxol-loaded micelles from block copolymers of PDLLA and PEG [186]. Films from multiblock copolymers of PLLA and PEG with fibroblast growth factor were investigated for wound healing applications [137]. Constant and long-acting release of levonorgestrel was obtained from PDLLA and P(LA-co-CL) microspheres [180]. Another steroid, progesterone, was encapsulated into PLLA microspheres [190] and microparticles from PDLLA containing progesterone were patented as a system for immunization of the internal female reproductive organs [65]. Ibuprofen, a non-steroidal anti-inflammatory drug, was released from PDLA microspheres in a biphasic fashion at a rate dependent on the particle size [191]. Macromolecular bioactive substances may also be encapsulated into and released from PLA matrices. The delivery of proteins from PLLA and PDLLA microspheres [192] and the delivery of DNA from PLLA microspheres have been presented [193].

3.1.3
Poly(lactide-co-glycolide), PLGA

Copolymerization of glycolide and lactide has been widely utilized to engineer the properties of PGA and PLA (PLLA or PDLLA) [19, 90, 144, 179, 194]. PLGA is less stiff than the original components, since the crystallinity decreases with an increase in the content of either comonomer. PLGA with compositions between 25 and 70% GA is amorphous [194], a desirable property in many drug delivery formulations. The hydrophilicity increases with increasing number of GA units in the polymer. PLGA has been in focus in the search for appropriate matrices for drug delivery microspheres, thanks to its ease of preparation, commercial availability at reasonable cost, versatility, biocompatibility, and hydrolytic degradation into resorbable, harm-less products [15, 90]. The popularity of PLGA is further explained by the fact that FDA has approved PLGA for a number of clinical applications.

After 25 years of huge and ever increasing interest in PLGA for drug delivery applications, the literature is extensive. However, the number of commercially available products on the market is not that large, for a number of reasons discussed in Sect. 1.1. One example of a commercially successful drug delivery system is Lupron Depot. This system is intended for endometriosis and prostatic cancer therapy and comprises one-month injectable microspheres of PLGA containing leuprorelin acetate (LH-RH agonist) [31, 32].

Non-steroidal anti-inflammatory drugs, e.g., diflunisal [103] and diclofenac sodium [104, 195], have been incorporated into PLGA microspheres and investigated for the treatment of rheumatoid arthritis, osteoarthritis, and related diseases. The encapsulation of biomacromolecules, e.g., proteins and vaccines, into polymeric microspheres presents a formidable problem because of the delicacy of these agents; bioactivity might be lost during preparation, and the release may be poor due to adsorption and/or aggregation. For instance, the release of recombinant human interferon-γ from PLGA microspheres was incomplete and the instability of the system limited its use to 7 days or less [196]. Similarly, incomplete release of lysozyme, recombinant human growth hormone, and a nerve growth factor from PLGA microspheres was reported [197, 198, 199]. Hence, much effort has been spent in evaluating PLGA delivery systems, with special regard to microsphere preparation, protein stability, and release characteristics [200]. Model proteins studied include bovine serum albumin, lysozyme, transferrin, and trypsin [99, 197, 201]. Several peptides, including vapreotide and rismorelin porcine, have been successfully incorporated and released from PLGA microspheres [202–204]. Systems for the controlled release of antigens have a great potential as vaccine adjuvants [66, 67, 201]. Recently, several studies of controlled release systems for DNA have been presented. DNA of different sizes has successfully been incorporated into PLGA microspheres but the loss of DNA integrity and activity still remains an important issue to be solved for these systems [193, 205–207].

Controlled release systems are highly interesting for cancer treatment, where improved treatment efficacy, site-specific administration, and reduced adverse side effects are desirable. Camptothecin [208] and taxol were both successfully incorporated into PLGA microspheres. The latter system contained isopropyl myristate and sustained release was obtained in vitro as well as in vivo in the lungs of mice [64]. There is also an increasing interest in the prolonged release of steroids, e.g., in the treatment of postmenopausal women or aging men. For this reason, the sustained release of β-estradiol from PLGA microspheres has been presented by several authors [209–211]. PLGA has also been investigated for the treatment of schizophrenia. Microspheres from PLGA released haliperidol over a period of 4 to 9 weeks depending on particle size [100]. Microspheres prepared from PLGA and PLGA/PCL blends were loaded with nerve-growth factors and ovalbulmin, intended for the treatment of central nervous system injuries [102]. Rods of PLGA containing disulfiram have been evaluated in rats as a prolonged delivery system for the treatment of alcohol addicts [212]. PLGA was also used as a component in a system intended for the prolonged release of doxycycline hyclate to periodontal pockets [213].

3.1.4
Poly(ε-caprolactone), PCL

Following the recognition of polylactide as a promising biomedical polymer, attention was drawn to related polyesters in the search for new degradable polymers in similar applications. PCL was recognized as a biodegradable and nontoxic material.

PCL is obtained by ring-opening polymerization of the 6-membered lactone, ε-caprolactone (ε-CL). Anionic, cationic, coordination, or radical polymerization routes are all applicable [106]. Recently, enzymatic catalyzed polymerization of ε-CL has been reported [172, 173]. PCL crystallizes readily due to the regular structure and has a melting temperature of 61 °C. It is tough and flexible [106]. The Tg of PCL is low (–60 °C). Thus, PCL is in the rubbery state and exhibits high permeability to low molecular species at body temperature. These properties, combined with documented biocompatibility, make PCL a promising candidate for controlled release applications [105]. PCL degradation proceeds through hydrolysis of backbone ester bonds as well as by enzymatic attack. Hence, PCL degrades under a range of conditions, biotically in soil, lake waters, sewage sludge, in vivo, and in compost, and abiotically in phosphate buffer solutions [214–217]. Hydrolysis of PCL yields 6-hydroxycaproic acid, an intermediate of the ω-oxidation, which enters the citric acid cycle and is completely metabolized. Hydrolysis, however, proceeds by homogeneous erosion at a much slower rate than PLA and PLGA [217]. Hydrolysis of PCL is faster at basic pH and higher temperatures [106].

Many approaches of varying the PCL properties have been described. Above all, there is an interest in accelerating the degradation rate. The range of PCL properties can be extended by copolymerization with many other lactones, such as glycolide, lactide, δ-valerolactone, ε-decalactone, poly(ethylene oxide), and alkyl-substituted ε-CL [20, 137, 144, 182, 218]. For instance, copolymers from ε-CL and L-LA were reported to degrade in vitro at a rate dependent on the L-LA content and PCL crystallinity. These copolymers were also biodegraded by microorganisms [219]. A biodegradable, cross-linked material based on PCL has been prepared from tetraethoxysilane end-capped PCL oligomers by the sol-gel process [220]. A valuable property of PCL is its remarkable compatibility with numerous other polymers [221]. Blends of PCL with other degradable polymers have a great potential for drug delivery applications. Degradable blends have been prepared from PCL and PHB, PLA or poly(mandelic acid) [141, 187]. PCL/PGLA blends have been used to prepare microspheres containing nerve growth factors for the treatment of central nerve system disorders [102].

Because PCL hydrolyzes slowly compared to PLA and PLGA, it is most suitable for long-term drug delivery. Capronor®, a 1-year contraceptive represents such a system [106]. The release of drugs is diffusion controlled rather than erosion controlled [106]. Nitrofurantoin, an antibacterial agent used in the treatment of urinary tract infections, has been incorporated into PCL microspheres [105]. The drug release rate was proportional to the square-root of time, i.e., following the Higuchi equation. The same mechanism was reported for the release of progesterone from PCL films [20]. Nifedipine (a calcium antagonist) and propranolol (a beta blocker) were incorporated into PCL microspheres as a hydrophobic and a hydrophilic model drug, respectively. Both drugs were satisfactory encapsulated and released from the matrix although the burst was higher and the encapsulation efficiency lower for the hydrophilic drug [222]. Levonorgestrel was released in a constant and long-acting fashion from P(LA-co-CL) microspheres [180]. The delivery of levamisole hydrochloride from PCL matrices could be manipulated by applying a coating of PLA or PLGA [223]. PCL has also been investigated for protein and peptide delivery, as the polymer barrier serves to protect the labile drugs from degradation in the gastrointestinal tract [63, 224]. Likewise, PCL was suggested as a delivery vehicle for recombinant growth hormone [225]. In the veterinary field, slow release systems for trypanocidal drugs have been studied [218]. PCL has been found less suitable for the release of ionic species [106].

As previously mentioned, degradable microspheres have gained attention as promising delivery vehicles for steroids in postmenopausal therapy. Copolymers of CL and D,L-LA were used to prepare microspheres for prolonged release of progesterone and β-estradiol. The system offered a constant release for up to 40 days in vitro and 70 days in vivo [226]. Similarly, PCL copolymers have been considered useful for androgen replacement therapy in the treatment of aging men with a testosterone deficiency. Micelles of PCL-block-poly(ethylene oxide) released dihydrotestosterone in a controlled fashion over 30 days. The biocompatibility was confirmed in vitro in a HeLa cell culture [227].

3.1.5
Poly(3-hydroxybutyrate), PHB and Other Poly(hydroxyalkanoate)s

In the 1920s, Lemoigne discovered a polyester, belonging to the group of poly(hydroxyalkanoate)s, in *Bacillus megaterium*. It was subsequently established that this polymer, poly(3-hydroxybutyrate) (PHB), is produced by a large number of bacteria, e.g., those present in soil or sewage, as an intracellular reserve of energy or carbon [228–230]. The physical properties of PHB are similar to those of polypropylene (PP), having a melting point of 175 °C and a Tg of –4 to +10 °C, depending on crystallinity [107, 109, 230]. PHB is, however, generally stiffer than PP and is optically active. Native intracellular PHB is amorphous but it crystallizes readily when isolated. To reduce the high crystallinity and resulting brittleness of PHB, it is often copolymerized with 3-hydroxyvaleric acid, HV. Copolymers of HB and up to 30% of HV are commercially available under the tradename Biopol. These materials were mainly developed at ICI [231]. PHB and its copolymers with HV are soluble in a number of common organic solvents, e.g., chloroform, acetic acid, and dimethylformamide [107].

PHB and its copolymers have attracted much attention because they are produced biosynthetically from renewable resources. The process involves fermentation of bacteria under controlled conditions, allowing for cellular overproduction and thus high yields [107, 230]. The polymeric granules can subsequently be extracted by organic solvents, such as chloroform [229]. The pathway of PHB biosynthesis have been widely investigated and shown to start with acetyl-coenzyme A and proceed to PHB via the polymerization of D-(-)3-hydroxybutyryl-CoA, mediated by PHB polymerase [230, 232, 233]. The exact mechanism of biosynthesis and the enzymes involved are, however, known to vary between microorganisms. Glucose is the most common substrate fed to the bacteria for commercial production but other substrates, e.g., sucrose, ethanol, alkanoic acids, or methanol, may be used as well [107]. By varying the carbon substrate, copolymers of HB with varying amounts of HV can be produced [231]. Doi et al. has shown that a number of different poly(hydroxyalkanoate)s, e.g., copolymers of HB and 4-hydroxybutyrate or 3-hydroxyhexanoate, may be produced by recombinant bacteria fed on plant oils or *n*-alkanoic acids [234]. Poly(hydroxyalkanoate)s with repeating units containing up to 12 carbons has been produced using various *n*-alkanoic acids as substrates [235].

PHB and P(HB-*co*-HV) are degraded by microorganisms such as bacteria, fungi, and algae under various conditions, e.g., soil, sewage, and sea water [110, 231, 236]. For example, biotic hydrolysis is catalyzed by several bacteria secreting active esterases, such as *Pseudomonas lemoignei* and *Alcaligenes faecalis* [35]. Enzymatic hydrolysis proceeds at the hydroxy end of the polymer chain, producing monomers, dimers, and trimers as degradation products. Furthermore, PHB and P(HB-*co*-HV) degrade under abiotic conditions by hydrolysis at a rate much slower that those of PLA and PLGA. Chemical hydrolysis proceeds by random chain scission of backbone ester groups [110]. The hydrolytic degradation rate increases with increasing pH and temperature and is dependent on the sample preparation technique, e.g., solvent casting or injection molding [107]. Mass loss of P(HB-*co*-HV) has been reported to approximate zero-order kinetics. This, combined with negligible water uptake, implies surface erosion [107]. Another study suggested that surface erosion only takes place in the initial stages of hydrolysis while the later stages are characterized by massive, rapid mass loss and increasing bulk porosity, typical of bulk erosion [110]. The slow degradation rate of PHB was reflected in insignificant changes of mechanical properties over a period of 6 months in vitro or subcutaneously in rats. Copolymerization with HV retarded the degradation rate at elevated temperatures [237]. Their degradability, combined with the fact that they are derived from renewable resources, have rendered PHB and P(HB-*co*-HV) highly interesting materials for packaging and disposable items.

Copolymerization offers considerable potential of engineering the properties and degradation rate of PHB over a broad range. HB has been copolymerized with a number of related species, e.g., ε-CL, LA, 4-hydroxybutyrate and hydroxypropionate. Furthermore, a number of PHB blends have been prepared. Miscibility of PHB with PCL, PLA, POE, poly(mandelic acid), and poly(sebacic anhydride) has been demonstrated. PHB-poly(sebacic anhydride) blends offered controlled release of bupivacaine at a rate dependent on the PHB content [141]. Blending of P(HB-*co*-HV) with natural polysaccharides was proposed as a way to manipulate the degradation rate of the matrix over a wide range [238]. Using PCL instead of polysaccharides resulted in blends more stable towards hydrolytic degradation while the initial tensile modulus was decreased. P(HB-*co*-HV) was, however, compatible only with low levels of PCL (10%) while higher levels of PCL caused phase separation to occur [239]. Blends of P(HB-*co*-HV), poly(ethylene adipate) and 20% PCL appeared incompatible and yielded a heterogeneous mixture of microporous, macroporous, and smooth particles when microcapsules were prepared by the double emulsion technique. These microcapsules released bovine serum albumin in a sigmoidal fashion over 25 days [240].

PHB and P(HB-*co*-HV) have several merits as matrices for controlled drug delivery [107, 109]. Their biosynthetic production excludes the use of solvents, initiators, or catalysts that could, if not properly removed from the biomedical device, pose a toxicological hazard to the patient. The materials are enzymatically as well as hydrolytically degradable. Biocompatibility does not seem to be a problem; the monomer D-(-)-3-hydroxybutyrate is in fact a normal constituent of blood [231]. The ease of crystallization of PHB during precipitation makes entrapment of the drug difficult [107]. Hence, copolymers with HV have been more popular for drug formulation than the pure PHB.

Microcapsules from PHB and P(HB-*co*-HV) has been prepared by various techniques and investigated for the release of bovine serum albumin [241]. PHB has also been suggested as a suitable matrix for drug delivery in veterinary medicine, for instance in the rumen of cattle [231]. Another study demonstrated that the microencapsulation technique, oil-in-water or oil-in-water-in-oil emulsion, has a considerable influence upon the morphology of PHB microparticles. A follicle stimulating hormone was successfully incorporated using the latter method [108]. Hydrolysis and erosion was slow, reported release times thus indicate that drug release is mediated by water penetration, formation of pores, and subsequent drug diffusion. PHB and P(HB-*co*-HV) are both crystalline materials [177] but the actual degree of crystallinity varies considerably depending on crystallization conditions and HV content [230]. This will, in turn, affect the drug release characteristics. Drug release rates have been reported to increase with increasing HV content [230, 242, 243]. For instance, microspheres from P(HB-*co*-HV) containing tetracycline for the treatment of periodontal diseases showed decreasing encapsulation efficiencies and increasing delivery rates with a higher HV content. The release behavior was better described by the Higuchi equation for zero-order kinetics, suggesting diffusion-mediated drug release. These results are in accordance with a study on microspheres from P(HB-*co*-HV) for the release of progesterone [242] showing that increasing contents of HV lead to higher porosity and hence higher delivery rates. Microspheres from P(HB-*co*-HV) has also been investigated for the delivery of anti-cancer drugs, such as aclarubin [244] or lastet [245]. The latter study showed that the release rate could be enhanced by incorporating glycerol tristearate as an additive. Another approach to cancer treatment was the incorporation of prodrugs of 5-fluoro-2′-deoxyuridine into PHB microspheres. These microspheres showed good biocompatibility in mice and rats [246].

3.2
Polyanhydrides

Polyanhydrides were first synthesized as early as 1909 [247] but it was not until the 1930s that their chemistry was more thoroughly explored. The formation of cyclic anhydrides from diacids in the HOOC-(CH_2)x-COOH series, their tendency to form rings, and their conversion upon heating into polymeric compounds were then systematically studied at DuPont [248]. The polyanhydrides were, however, regarded as being of minor importance since they are hydrolytically unstable and degradable. Fifty years later, these very properties made the polyanhydrides excellent candidates for drug delivery applications. Aliphatic polyanhydrides degrade within days or weeks while the erosion of aromatic polyanhydrides ranges from several months to years. The erosion time can thus be varied over a broad range by changes in the polymer backbone [249]. Furthermore, polyanhydrides are potentially surface eroding. Many polyanhydrides have fairly low melting points and are soluble in common organic solvents. These properties, together with their biocompatibility [112, 250], make polyanhydrides popular candidates for controlled release applications. A broad variety of polyanhydrides has been presented over the years, based on one or a combination of aliphatic, aromatic, and heterocyclic monomers (Fig. 6).

The classical route for preparing polyanhydrides is by melt polycondensation. A dicarboxylic acid is refluxed with an excess of acetic anhydride yielding a solution of mixed anhydride. The excess acetic anhydride is subsequently removed in vacuum at high temperatures, resulting in high polyanhydrides. Polycondensation reactions of diacids with acetic anhydride, acid chlorides, or phosgene, also yield polyanhydrides. The formation of polyanhydride from diacids in acetic anhydride is an equilibrium reaction, limiting the final yield and molecular weight. To circumvent the equilibrium, a synthesis route for aliphatic polyanhydrides from mixed anhydrides of diacids and ketene was developed [251], focusing on the preparation and characterization of poly(adipic anhydride) (PAA) [252]. Ring-opening polymerization of cyclic anhydrides presents a cleaner and more simple route to high polyanhydrides [253, 254]. For instance, PAA may be synthesized via ROP of the cyclic monomer oxepan-2,7-dione in melt, solution or bulk [255–257], as shown in Fig. 7.

The highest molecular weight polyanhydrides have been obtained by melt condensation. The reaction temperature and time, the presence of catalysts, and the purity are important factors in this respect [253].

Pharmaceutical research has to date been focused on polyanhydrides derived from sebacic acid (SA) and its copolymers with bis(*p*-carboxyphenoxy)propane (CPP) [75, 113, 115, 119]. More recently, a new class of polyanhydrides was presented, containing fatty acid dimers (FAD) [116, 118, 258]. Erosion characteristics, microsphere preparation, pH-dependence, release rates, morphology, and in vivo performance of polyanhydrides from SA, CPP, and FAD have been intensely studied [75, 111–115, 117, 119, 258–260]. Other unsaturated polyanhydrides have been derived from ricinoleic acid [261] and ricinoleic acid half-es-

Degradable Polymer Microspheres for Controlled Drug Delivery

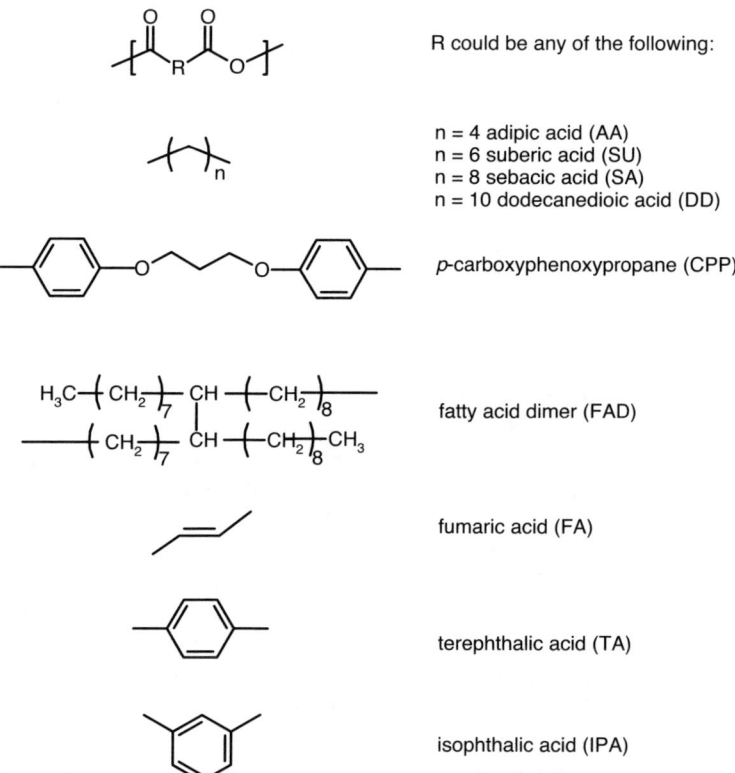

Fig. 6. The molecular structure of some polyanhydrides

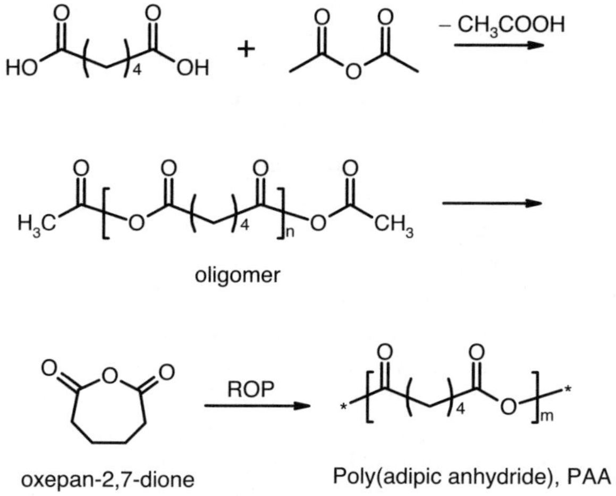

Fig. 7. Synthesis of poly(adipic anhydride)

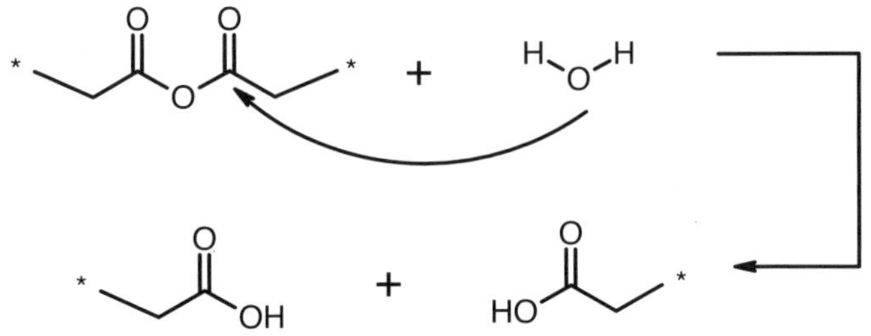

Fig. 8. Schematic hydrolytic degradation of an aliphatic polyanhydride

ters with maleic and succinic anhydrides [262]. The fatty acid building blocks enhance the hydrophobicity of the matrix, preventing water penetration and making the matrices more surface eroding. Recent developments in polyanhydride chemistry include polyanhydrides made from bile acids [263] or polyanhydrides containing imide groups [264, 265] or poly(ethylene glycol) (PEG) segments [266]. The introduction of imide groups adds mechanical strength to the material while PEG enhances the hydrophilicity and rate of degradation.

Polyanhydride hydrolysis proceeds via free carboxylic acid chain-ends as outlined in Fig. 8 to yield carboxylic acids as final degradation products. The anhydride bond is more susceptible to hydrolytic cleavage than the ester and carbonate linkages [249]. Depending on the polymer structure between anhydride bonds, matrix life times of hours to many months may be achieved. PAA is a relatively hydrophilic polyanhydride due to the short distance between the labile backbone linkages and it degrades rapidly; PAA of low molecular weight is totally degraded within two days [257]. The generation of carboxylic functional groups may, if erosion is slower than degradation, result in an acidification of the microenvironment inside a degrading polyanhydride device. Since the hydrolysis of polyanhydrides is retarded at lower pH, the erosion may proceed faster at the surface than in the bulk [42, 253]. Matrices with labile polyanhydride bonds and hydrophobic backbones are thus potentially surface eroding and may offer close to zero-order release of encapsulated drug [75, 111].

Polyanhydride devices for controlled delivery of local anesthetics [114] and chemotherapeutic agents [115] have been investigated. Rapidly degrading PAA was investigated as a promising matrix for ocular drug delivery [267]. A few drug delivery systems based on polyanhydrides are currently being tested for clinical use. One example is the GLIADEL wafer; a P(CPP-co-SA) (20:80) matrix containing the chemotherapeutic agent BCNU for brain cancer treatment [268]. The FDA has approved P(CPP-co-SA) for this application. Microspheres from P(CPP-co-SA) loaded with acetylcholine have also been studied for local delivery to the brain [269]. The use of polyanhydrides in oral delivery of insulin and plasmid DNA has been investigated [117]. Recently, a polyanhydride from erucic acid dimer (EAD) and SA was investigated for the delivery of heparin. Controlled release was observed over 20 days when the matrices were coated with PLA

[270]. Poly(EAD-*co*-SA) was also used for the sustained release of gentamicin, intended for the treatment of osteomyelitis [271]. P(FA-*co*-SA) matrices have been appreciated for their bioadhesive properties, which serve to prolong the gastrointestinal residence time of oral delivery systems. Microspheres from P(FA-*co*-SA) were studied for the release of an anticoagulant drug, dicumarol [118]. SA has also been linked by ester bonds to salicylic acid to form a poly(anhydride-ester), which degrades by hydrolysis to directly release this anti-inflammatory analgesic [272]. The merit of this prodrug lies in the fact that the active moiety is incorporated in a biodegradable polymer backbone and not as a side-group on a non-hydrolyzable polymer chain. A poly(anhydride-*co*-imide) with L-tyrosine moieties in the polymer backbone was specifically prepared for vaccine delivery and used for the preparation of microspheres capable of controlled release of a model protein [264]. The anhydride functionality is potentially reactive toward nucleophiles such as amines. This must be kept in mind when considering the use of polyanhydrides for the delivery of amino-containing drug, e.g., peptides.

3.3
Aliphatic Polycarbonates

Aliphatic polycarbonates with methylene segments of n=2–14 and 18 were obtained from a homologous series of glycol esters of carbonic acid as early as the 1930s [273]. In recent years, aliphatic polycarbonates have been explored in the search and design of new polyester-related structures for medical applications. Ring-opening polymerization of cyclic carbonates presents a smooth route of polymerization, particularly for substituted and unsubstituted six-membered rings [274, 275]. Both anionic and cationic polymerization has been described. Enzymatic ROP presents an alternative, cleaner route of polymerizing cyclic carbonates. Using porcine pancreatic lipase as the catalyst, high molecular weight poly(trimethylene carbonate) (PTMC) was prepared in bulk at 60–100 °C for 24 h [276]. From seven lipases studied, a lipase from *Candida antarctica* was identified as the most effective catalyst for ROP of trimethylene carbonate with respect to molecular weight and monomer conversion [277]. In recent years, there has been a growing interest in polymers that can be obtained from renewable resources. 1,3-Propanediol represents an interesting monomer in this respect as it can be obtained by microbial fermentation of glycerol. The reaction of 1,3-propanediol with succinic acid (obtained by fermentation of starch) yields biodegradable poly(ester carbonates) having a high potential as environmentally friendly polymers [278].

Among the polycarbonates, PTMC in particular has been studied for biomedical use. PTMC is commonly derived by ring-opening polymerization (Fig. 9). The monomer, 1,3-dioxan-2-one, is obtained by refluxing 1,3-propanediol in an excess of diethyl carbonate. Albertsson et al. have reported on the polymerization in bulk and solution using a number of initiators and catalysts [120, 279, 280]. PTMC is aliphatic and has a low Tg (approximately –17 °C), it is thus in its

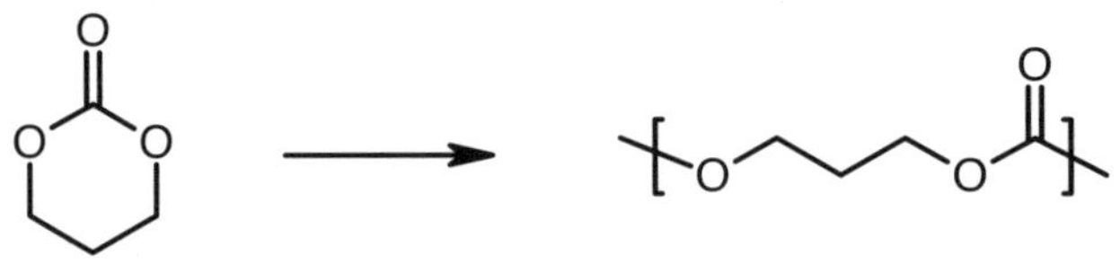

Fig. 9. Poly(trimethylene carbonate), PTMC

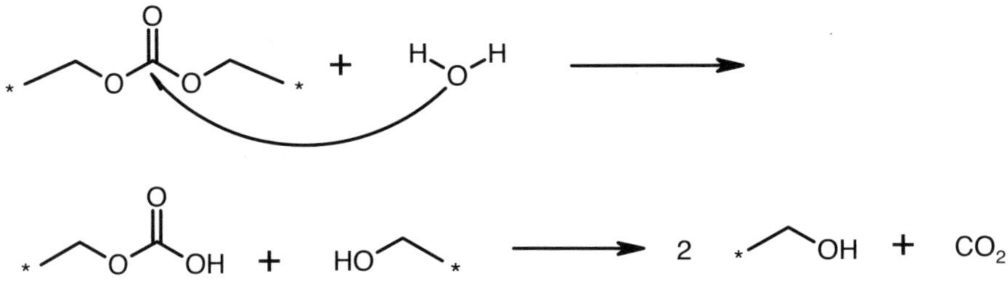

Fig. 10. Schematic hydrolytic degradation of an aliphatic polycarbonate

rubbery state a room temperature. It may appear as anything from a sticky and rubbery consistency to a hard solid, depending on the molecular weight [121]. Like many aliphatic polycarbonates, PTMC displays significant softening at 40–60 °C [177]. This allows for easy processing and for encapsulation of sensitive drugs under mild conditions. A family of aromatic polycarbonates derived from tyrosine has recently been presented as promising biomedical materials which may lend themselves to drug delivery or bone fixation applications [281].

PTMC degrades by hydrolysis and by enzymatic promotion. Thermal degradation occurs at elevated temperatures (190 °C) [282]. The polycarbonate bond is intrinsically more sensitive to hydrolysis than the polyester linkage [120]. Nevertheless, PTMC hydrolyses very slowly, properties of high molecular weight PTMC may be unaffected for more than a year in vitro [257]. The molecular weight decrease of PTMC is very slow in aqueous solution and furthermore independent of the nature of the aqueous medium. Complete hydrolysis can take years. The slow hydrolysis rate of PTMC has been attributed to the lack of an autocatalytic process [120]. Hydrolysis finally yields diols and carbon dioxide, as illustrated in Fig. 10. The degradation products of PTMC have been identified by gas chromatography-mass spectrometry analysis [257]. 1,3-Propanediol was found to be the major degradation product of PTMC, which is in accordance with the degradation mechanism proposed in Fig. 10.

Copolymerization of TMC with ε-CL or L-lactide has been reported to be useful for modifications of the polymer properties [144, 218, 280, 283]. The molecular architecture presents a powerful tool for obtaining new materials with interesting properties. For instance, a star-shaped rubbery poly(TMC-co-CL) was synthesized. D,L-LA/GA polymerization was then initiated from the hydroxy terminated arms to yield a poly(TMC-co-CL)-block-PLGA [284].

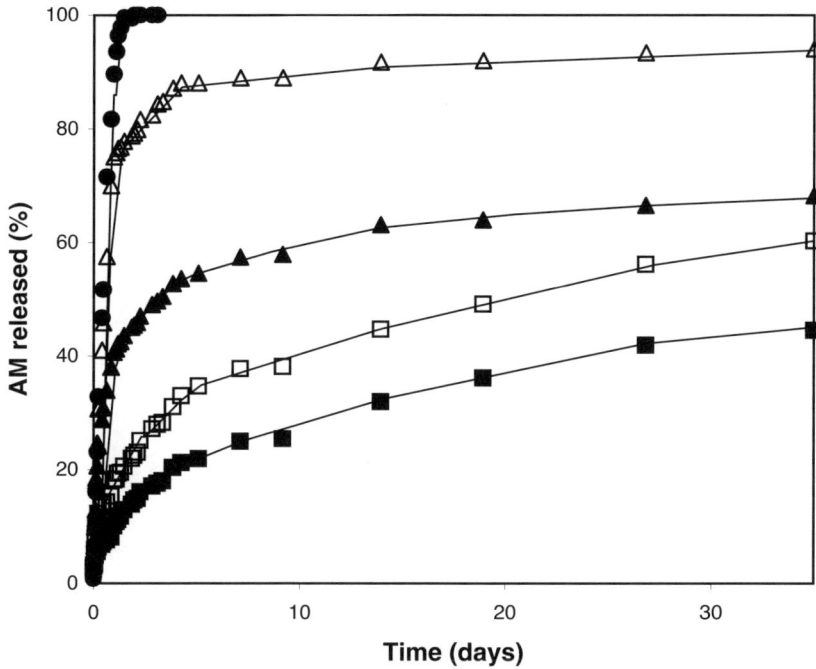

Fig. 11. Release of drug from poly(trimethylene carbonate)-poly(adipic anhydride) (PTMC-PAA) blends with different amounts of PAA: (●) 100%, (□) 80%, (▲) 50%, (□) 20%, and (■) 0%

Blending is yet another way of manipulating the polymer properties. PTMC degrades too slowly to be useful for most drug delivery applications. Blends were prepared in an attempt to enhance and control the drug delivery rate [121]. Films were prepared from PTMC and poly(adipic anhydride) (PAA), which were partially miscible and formed macroscopically homogeneous blends. The PAA was found to act as a plasticizer and facilitated the erosion of PTMC by increasing the porosity and hydration. PTMC-PAA blends offered sustained and controllable release of an incorporated therapeutic substance as shown in Fig. 11 [121].

3.4
Poly(orthoesters), POE

In the early 1970s, the ALZA Corporation began its search for polymers suitable for erodible drug delivery systems. The ideal polymer was identified as one undergoing surface erosion in vivo and degrading to non-toxic, low molecular weight products at a rate that could be manipulated over a broad time span. To meet these criteria, a novel family of hydrolyzable polymers was developed, the poly(orthoesters), POEs [285]. The general structure is schematically shown in

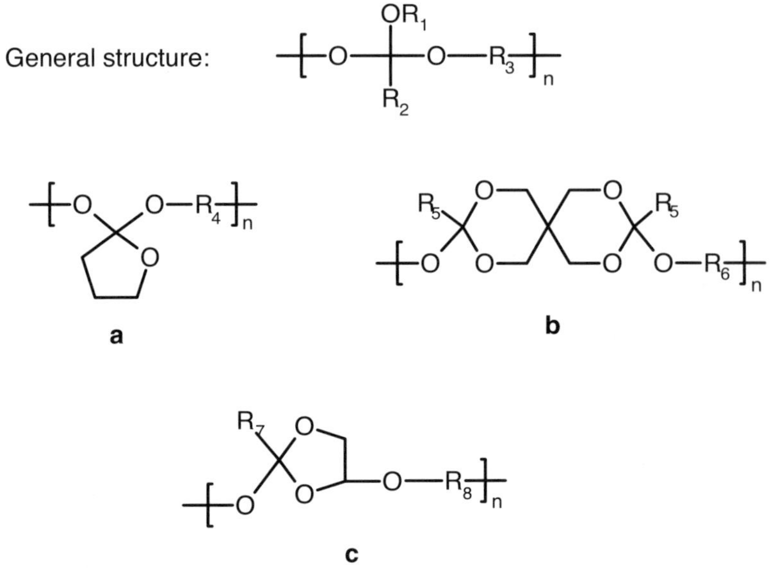

Fig. 12. The molecular structure of different classes of poly(orthoesters)

Fig. 12. POEs have several merits as matrices for drug delivery. The orthoester linkage is pH sensitive, rendering the polymer rather stable at neutral pH with the degradation rate slowly increasing as the pH of the surrounding medium decreases. As a result, the degradation rate can be manipulated to range from hours to months by the incorporation of excipients of acidic or basic nature [122]. By careful formulation, it is possible to design systems in which erosion is confined to the surface of the device.

The first poly(orthoesters) were developed at ALZA Corporation and are described in a series of patents [286–288]. They were prepared by transesterification of a diol and diethoxytetrahydrofuran (Fig. 12, *a*) and go under the tradename Alzamer (previously Chronomer). Initial development work showed that monomers with three active alkoxy groups would inevitably result in crosslinked products. Hence, the reactivity of one alkoxy group had to be suppressed by using cyclic monomers in order to achieve a linear POE. Upon hydrolysis of these polymers the diol is regenerated and γ-butyrolactone is formed; the latter readily hydrolyzes to hydroxybutyric acid [289]. The formation of an acidic degradation product will create an acidic microclimate inside the device and autocatalyze further degradation of the acid-labile POE. Hence, a basic excipient had to be incorporated to avoid massive bulk degradation of the device. Another drawback of this system was the tedious preparation procedure. Alzamer was investigated for the sustained delivery of several drugs, including naltrexone and the contraceptive steroids northindrone and levonorgestrel. Also, it was investigated for the treatment of burns. However, human trails showed that the steroid implant could cause local tissue irritation. Due to the drawbacks of this system, the development work was discontinued [289].

A class of new and improved POEs were developed by SRI International (Fig. 12, b) [122, 290]. These polymers were based on the addition of diols to diketene acetals. However, the preparation of linear polymers from linear diketene acetals was unsuccessful, since the reactivity of the alkoxy groups resulted in cross-linked products. Again, cyclic monomers were the solution to this problem. The addition of a diol to cyclic diketene acetals, typically 3,9-bis(ethylidene-2,4,8,10-tetraoxaspiro[5,5]undecane ($R_5=CH_2CH_3$ in Fig. 12, b), readily yields POEs using acid catalysts, such as p-toluenesulfonic acid [122]. This POE degrades by hydrolysis to form the monomeric diol and pentaerythriol dipropionate. The latter compound is often eroded from the device before it slowly hydrolyzes further to pentaerythriol and propionic acid. Hence, the risk of massive bulk erosion due to autocatalysis is eliminated [289]. Mechanical and physical properties of these polymers can be manipulated by careful selection of the diol or by adding a mixture of diols. While diols result in linear POEs, cross-linked polymers may be obtained by adding alcohols of higher functionality. Dense cross-linked matrices can be obtained because the polycondensation reaction proceeds without the generation of volatile by-products [122] A cross-linked material is prepared in two steps. First, a prepolymer is prepared from two equivalents of diketene acetal and one equivalent of diol. A network is then formed by reaction the prepolymer with triols or a mixture of diols and triols. Because the prepolymer is a viscous liquid, it can be mixed with drug substances and cross-linked at rather low temperatures (40 °C). Another interesting property for drug delivery applications is the possibility to control the degradation rate [123]. The use of acidic excipients, typically anhydrides, accelerates the erosion, while the erosion rate was retarded when basic excipients, such as $Mg(OH)_2$, were incorporated. Surface erosion and zero-order release from anhydride-containing POE disks was demonstrated [123]. The delivery of various drug has been studied, e.g., the narcotic antagonist naltrexone pamoate [124] and 5-fluorouracil, intended for tumor inhibition in cancer therapy [291]. A pH-sensitive POE was prepared from and N-methyldiethanolamine and 3,9-bis(ethylidene-2,4,8,10-tetraoxaspiro[5,5]undecane), intended for the pulsatile release of insulin [125].

More recently, a new class of POEs has been prepared by the reaction between a triol and an alkyl orthoacetate (Fig. 12, c) [126]. Depending on the triol used, everything from a sticky, ointment-like polymer to a solid, rigid material can be prepared. The use of 1,2,6-hexanetriol will produce erodible polymers with highly flexible backbones. Their consistency at room temperature is that of a viscous paste, allowing for therapeutic substances to be incorporated without the use of solvent or elevated temperatures. Such a matrix, containing 4-homosulphanilamide, was suggested for the treatment of burns [126]. Tetracycline was incorporated into a semisolid POE system for the treatment of periodontal diseases. Sustained release was obtained for a period of days to several weeks, and could be controlled by adding various amounts of a basic excipient [$Mg(OH)_2$]

[127]. The biocompatibility of an ointment-like POE was investigated in vivo in rabbit eyes [83]. The subcunjuctival injection triggered an acute inflammatory response which was attributed to the formation of acidic by-products. Recently, a POE prepared from 1,2,6-hexanetriol and trimethyl orthoacetate was investigated for the delivery of 5-fluorouracil and mitomycin C in glaucoma filtering surgery [128, 292]. These POEs are partially degraded upon γ-irradiation and the biocompatibility could be improved by avoiding this sterilization procedure [128].

3.5
Poly(1,5-dioxepan-2-one), PDXO

The aliphatic poly(ether lactone)s are a group of synthetic polymers with high elasticity and high tissue absorptivity [293]. The ether function in the polymer backbone adds flexibility to the ester chain. Ring-opening polymerization of 1,4-dioxan-2-one yields an elastic polymer, polydioxanone, with a tensile strength similar to that of human tissue [294]. Polydioxanone has been successfully used to prepare monofilament sutures, with a flexibility superior to that of PGA sutures [294]. Recently, the lipase-catalyzed polymerization of polydioxanone was demonstrated [295].

Poly(1,5-dioxepan-2-one) (PDXO) (Fig. 13) is derived from the seven-membered analogue of the ether lactones, 1,5-dioxepan-2-one (DXO). DXO was developed and copolymerized with glycolide or D,L-lactide to obtain more flexible biomaterials [296, 297]. However, the suggested synthetic methods gave low yields and new and improved synthetic routes for the preparation of DXO and the ring-opening polymerization of this monomer were introduced by Albertsson and Mathisen in 1989 [153, 298]. Since then, much research has been undertaken to develop synthetic pathways for homo- and copolymers of DXO with high molecular weight at high yield [150, 153–156, 293, 299, 300]. In spite of the regular structure, PDXO does not crystallize but is totally amorphous with a Tg

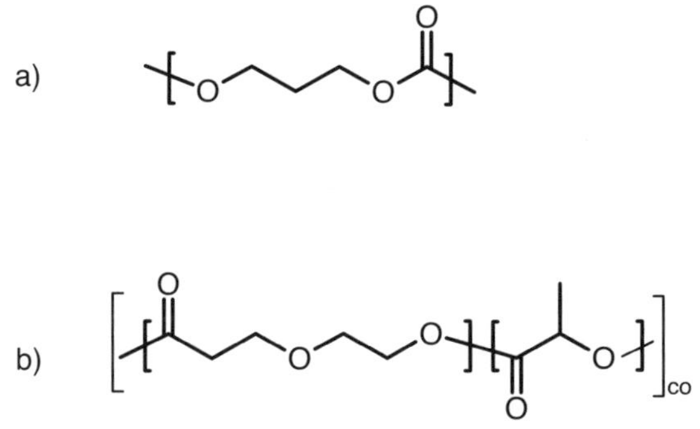

Fig. 13. *a*) Poly(1,5-dioxepan-2-one) (PDXO) and *b*) a random copolymer of DXO and lactide, P(LA-*co*-DXO)

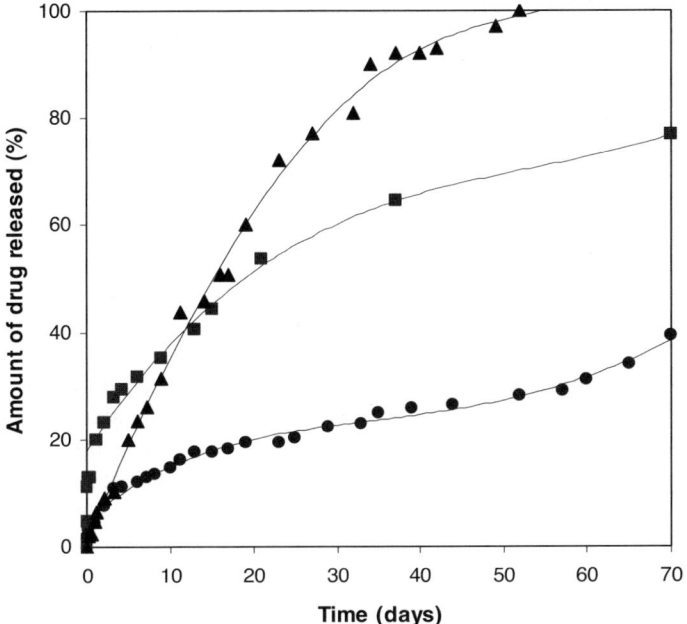

Fig. 14. Release of a therapeutic substance from (▲▲) P(L-LA-co-DXO), (■■) PDLLA-PDXO, and (●●) PLLA-PDXO microspheres with a L-LA/DXO molar ratio of 90:10

of −36 °C. Due to the low Tg, it appears as a sticky and gel-like material at room temperature. PDXO has poor mechanical properties but can be cross-linked to form degradable elastomers [301]. PDXO is hydrolytically degradable [129, 293]. Degradation products of P(L-LA-co-DXO) have been identified by headspace gas chromatography-ion trap mass spectrometry [302]. Lactic acid and 2-hydroxyethoxypropanoic acid were found to be the major degradation products.

By introducing DXO units into the backbone of PLA, the Tg of the lactides may be lowered to or below body temperature, and the crystallinity may be lowered. Thus, more flexible materials are obtained where the properties and degradation rate can be varied over a broad range by means of the composition. This is interesting for biomedical applications. Blends of PLA and PDXO provide additional opportunities to vary the properties. There is a large difference in reactivity ratio of LA and DXO [154]. As a result, a more block-like copolymer structure than expected for a totally random copolymer is obtained, even though the structure is somewhat randomized by transesterification reactions. In contrast, DXO and ε-CL form a truly random copolymer [156]. Tri-block copolymers of LA and DXO have recently been described [303].

Microspheres were prepared from copolymers of DXO and L-LA and from homopolymer blends of DXO with L-LA and D,L-LA [129–131]. PLLA was partially miscible with PDXO and formed semicrystalline and dense microspheres. PDLLA and PDXO were fully miscible and formed homogeneous and amorphous microspheres. Morphological differences between copolymers and blends had a significant effect on the degradation and release rates. Sustained release of the incorporated drugs was obtained from all matrices as exemplified in Fig. 14. The release and degradation rates can be manipulated by means of blend or copolymer composition [129–131].

4
Microspheres

4.1
General

The technology of microencapsulation, the packaging of liquids and solids in spherical particles of micron size, emerged in the mid 1950s and has been used for various applications, including graphic products, optics, agricultural chemicals, adhesives, perfumes, food and flavorings [304]. Interest in microencapsulation for the formulation of drug delivery devices has accelerated in the last decade.

4.2
Preparation

A number of methods has been presented for the preparation of drug-containing microspheres from biodegradable polymers [68, 305–307]. The first and previously widely used technique is coacervation. Coacervation, or phase separation, involves the dissolution of the polymer in a liquid in which the insoluble core material to be encapsulated is suspended [308]. Compositional changes of the system, e.g., addition of salts, or a pH or temperature change, subsequently bring about precipitation of the polymer. Particles manufactured by this method are capsular in structure. Coacervation may thus be used for the entrapment of liquids and oils. The National Cash Register Co. performed the pioneering work on coacervation [308].

Oil-in-water (O/W) solvent-evaporation techniques [68, 305, 309], schematically presented in Fig. 15, are amongst the most popular microencapsulation methods due to their relative ease of processing and versatility. The therapeutic substance and the polymer are dissolved in a volatile organic solvent, typically methylene chloride (CH_2Cl_2). This oil phase is then dropwise added to a water phase, the latter containing a stabilizer such as polyvinyl alcohol or gelatin, under vigorous stirring. The immiscibility of the two phases allows the formation of a stable emulsion. The role of the stabilizer, also referred to as emulsifier, is to prevent the droplets from coalescence and coagulation so that a stable emulsion is preserved. The polymer precipitates as the solvent is removed by evaporation, a process sometimes facilitated by reduced pressure or by the addition of a nonsolvent. The hardened microspheres are subsequently separated from the aqueous phase, washed, and dried.

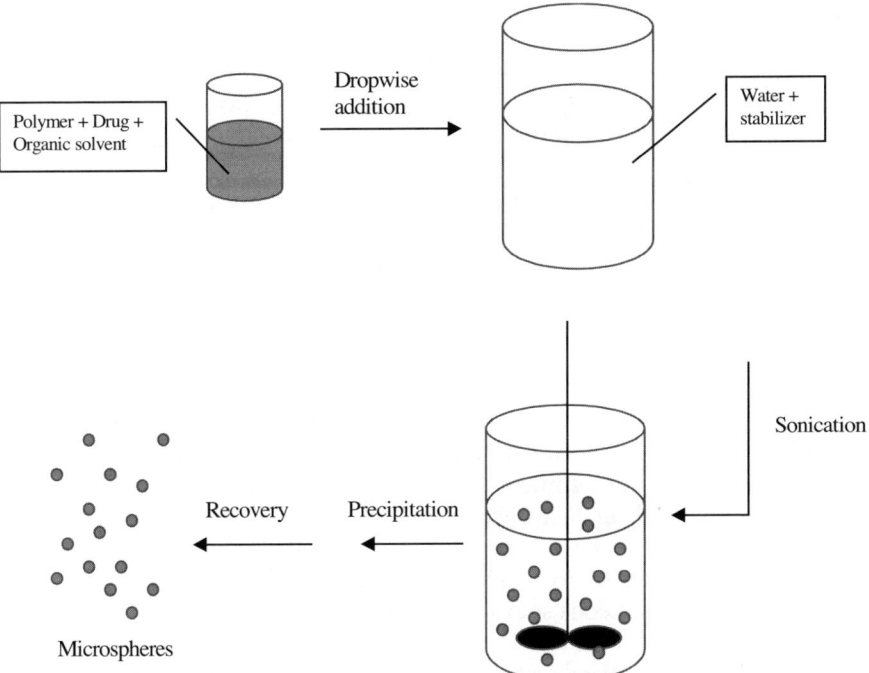

Fig. 15. Schematic illustration of microsphere preparation by oil-in-water (O/W) solvent evaporation technique

When selecting a dispersed phase solvent, the following factors must be considered [305]:
- Ability to dissolve the polymer;
- Immiscibility with the continuous phase;
- High volatility;
- Low toxicity.

The solvents commonly used for o/w solvent evaporation, CH_2Cl_2 and acetonitrile, amply meet the first three criteria. They are, however, toxic and special care must be taken to remove all traces of solvent at the end of the manufacturing process.

The continuous phase should meet the following demands [305]:
- Inability to dissolve the polymer;
- Immiscibility with the dispersed phase;
- Lower volatility than the dispersed phase solvent;
- Low toxicity;
- Easy separation from the precipitated microspheres.

Distilled water is often used as a hydrophilic continuous phase and meets these criteria well, biocompatibility included.

The processing temperature has a great impact upon the microsphere morphology. At elevated temperatures, the rate of precipitation is too fast and the solvents chosen would "boil out" of the emulsion droplets. This is known to result in disrupted, distorted, or highly porous spheres [68, 307]. "Boiling out" may be prevented by conducting sphere preparation at low temperature or by applying pressure to the vessel. The type and concentration of the emulsifier may affect the microsphere size, shape, porosity, and encapsulation efficiency [68, 217]. The viscosity of the organic phase has also been reported to be an important factor affecting the surface morphology of the resulting microspheres [181]. The complexity of the sphere preparation technique is such that it is very difficult to predict method conditions for a specific polymer. In practice, optimal conditions have to be empirically found for each polymer-drug system to be formulated.

Low entrapment efficiency and high burst of drug upon incubation are problems often associated with the solvent-evaporation techniques when they are used for the preparation of microspheres containing water-soluble substances. This is explained by drug diffusion into the aqueous processing medium during preparation [68, 240, 305]. For the entrapment of water-soluble drugs, a double-emulsion solvent evaporation technique, water-in-oil-in-water (W/O/W), has been developed. An aqueous solution of the therapeutic substance is emulsified in an organic polymer solution. This water-in-oil emulsion is further emulsified into a continuous stabilized water phase. As the organic solvent is removed by evaporation, the polymer hardens and forms microspheres. This (W/O/W) encapsulation technique has been employed for the encapsulation of bioactive macromolecules like proteins, vaccines, and peptides. The technique is also referred to as in-water drying.

The techniques described above are not appropriate when dealing with polymers prone to hydrolysis, e.g., polyanhydrides, as water-containing emulsions may induce degradation and marked molecular weight loss [305]. Oil-in-oil (O/O) or water-in-oil-in-oil (W/O/O) solvent-removal techniques present useful alternatives [113]. The use of oil as the continuous phase prevents the drug from partitioning out during processing. The (O/O) method is quite similar to the (O/W) technique but involves the dissolution of polymer and drug in an organic solvent, typically methylene chloride, and the emulsification of this organic phase in a second stabilized oil phase. When the (W/O/O) technique is employed, an aqueous solution of drug is first emulsified in the organic polymer solution before the emulsion is added to the stabilized oil phase. Vegetable oils are preferred, since they are hydrophobic and edible. A stable emulsion is formed under vigorous stirring. Microparticles are formed as the organic solvent is extracted into the oil phase. The (W/O/O) technique, also termed in-oil drying, has been used for the encapsulation of various kinds of proteins [310].

When emulsion techniques are used for microsphere preparation, a number of processing factors influences the final structure of the microspheres, e.g., the choice of solvents and surfactants, phase viscosity, the ratio of the dispersed to the continuous phase, mixing speed, processing temperature, and time. Microspheres can be prepared over a wide size range, from ten to several hundred microns, by careful modification of processing conditions [68, 305, 306]. The surfactants investigated for emulsion techniques include gelatin, poly(vinyl alcohol), Tweens (polyoxyethylene derivatives of sorbitan fatty esters), alginates, methyl cellulose, poly(vinylpyrrolidone), PEG, PEG-dextrin conjugates, and PEG-PPO copolymers [68, 181, 305, 306].

Other techniques include spray drying [192, 202] and hot-melt techniques [224,249]. Spray-drying is the preferred technique for industrial sphere preparation. A polymer solution is sprayed into a heated chamber, which permits solvent evaporation and polymer precipitation. The method offers the advantage of allowing considerable control of the particle size. Only a limited number of substances can be incorporated by hot-melt techniques, since the drug must withstand high temperatures without loss of biological activity. A microencapsulation process based on fluidized bed coating has been reported [311]. This process involves the dissolution of the drug and the polymer in a mutual solvent. Microcapsules are formed as this solution is processed through a Wurster air suspension coater apparatus.

The encapsulation of bioactive agents, such as proteins or peptides, presents a special problem due to the delicacy of their structural conformation. Their biological activity may be irreversibly disturbed even by small changes of pH, temperature, or ionic concentration [192,200]. Spray-drying and double emulsion solvent evaporation techniques (w/o/w) have been suggested for the incorporation of proteins with retention of their bioactivity [99, 192, 202, 198].

4.3
Controlled Release Applications

Interest in microencapsulation for the formulation of drug delivery devices has accelerated during the last decade. When injectable microspheres are used, the inconvenient insertion of large implants is avoided. Microspheres larger than 6 μm should not be intravenously administered, since they cause acute toxicity reactions associated with capillary blockage in the pulmonary circulation [64]. Instead, microspheres are primarily intended for intramuscular or subcutaneous implanting. Alternatively, suspension in a suitable vehicle/thickening agent, such as a methylcellulose solution, may precede administration to regulate the consistency of the formulation [200]. Bioadhesive microspheres have a potential as oral delivery systems [118].

5
Sterilization and Storage

The commercial success of a drug delivery device is intimately connected to its ability to withstand sterilization and upon its storage stability prior to use.

5.1
Sterilization

The ultimate goal of sterilization is to destroy and remove all microorganisms from the medical device. In theory, the probability of survival decreases exponentially with time to which the microorganisms are subjected to a sterilizing agent. A survivor level or sterility assurance level (SAL) of 10^{-6} is generally required in Europe if the device is to be labeled "sterile" [312]. The traditional methods of sterilization of medical devices include the use of dry or moist heat, chemicals (ethylene oxide), or radiation. Currently practiced sterilization methods are not, however, well suited to biodegradable polymers [313]. The thermoplastic nature and the hydrolytic biodegradability of most biomedical polymers make them heat and moisture sensitive. The selection of a proper sterilization technique is therefore of utmost importance to the physicochemical and mechanical properties of the drug delivery system and hence to the performance in vivo.

Steam sterilization by autoclaving at 121 °C is the most widely employed method today but it may induce hydrolysis and/or melting of the polymer matrix and is, in effect, not recommendable for degradable polymers [313, 314]. Sterilization by dry heat requires temperatures of 160–170 °C for two hours or more. These conditions may cause melting, distortion, and/or degradation of the polymer and the technique is thus of limited use for most polymeric materials.

Chemical sterilization with ethylene oxide gas is a slow and fairly expensive method, but it offers the advantage of effective treatment at ambient temperature and is useful for hydrolytically unstable polymers [8]. Nevertheless, its popularity is decreasing due to the well-known toxicity and flammability of ethylene oxide. Since it is a strong alkylating agent, any gas residues may react with functional groups on the polymer surface, and thereby alter its biological properties. To ensure the complete removal of trace residues before use, the specimens must be thoroughly degassed. The effectiveness of EtOx sterilization is dependent on several parameters, e.g., the temperature, spore environment, level of vacuum, EtOx concentration, degassing procedure, and humidity. To ensure effective microbial death, a relative humidity above 30% is required [312]. Generally, the relative humidity is kept at 50–60% during the sterilization procedure and these conditions may cause an onset of degradation of hydrolytically labile polymers, which, in turn, may be detrimental to the performance and lifetime of the drug delivery device. Degradation may also occur during the degassing procedure, which involves storage at elevated temperature in air circulation chambers for 2–5 days, depending on the device porosity [313].

High-energy radiation sterilization has the advantages of high efficiency and negligible thermal effects. Radiation sterilization is desirable from an economic point of view and allows for packaging prior to treatment [8]. The major drawback of radiation sterilization is that many polymers undergo property changes induced by irradiation [314]. Generally, polymers exhibiting high heats of polymerization tend to cross-link upon radiation, with an apparent increase in mechanical stability with increasing radiation dose [313, 315, 316]. Polymers with low heats of polymerization tend to degrade by random chain scission rather than to cross-link upon radiation [313, 315, 316, 317]. This may affect the physicochemical properties. Radiation of polymers may also lead to double-bond formation, gas evolution, additive leakage, and discoloration [315]. The radiation sterilization of biodegradable polymer matrices may induce degradation reactions that may continue upon storage, affecting the erosion and drug release performance [313,315]. Small changes in material properties may be acceptable in short-term biomedical applications, while similar changes may have severe consequences in long-term applications [315]. The susceptibility to radiation of the incorporated therapeutic agent must also be considered. The geometry of the product and the product density are key factors that need consideration to ensure that the device receives an appropriate and uniform dose. An average dose of 25 kGy is acceptable for most medical devices [314,316]. γ-Irradiation was reported to cause deterioration of PGA and PLGA, evident as a decrease in molecular weight and a reduction in tensile strength following implantation [179]. An unzipping degradation mechanism was proposed. Other studies suggested that random chain scission was the predominating mechanism for radiation-induced degradation of PLGA [317]. PDLLA showed substantial embrittlement and a reduction of tensile strength upon γ-irradiation [317]. The influence of sterilization on the physicochemical properties of polyanhydrides is described much less frequently in the literature than its influence on polyesters. In a study of γ-irradiated poly(sebacic acid) and its copolymers with fatty acid dimer, no significant changes in ^{1}H-NMR spectra, molecular weight, color, or pliability were observed [253]. However, free radicals were detected in all samples of γ-irradiated polyanhydrides [318]. The stability of PTMC-PAA implants toward sterilization by β-irradiation or EtOx treatment was found to depend on the blend composition, as the polycarbonate component was much more stable that the hydrolytically sensitive polyanhydride [82]. γ-Irradiation of PHB reduced the tensile strength but did not significantly alter the elastic properties. The in vivo degradation rate was enhanced with irradiated samples compared to non-irradiated PHB [237]. Polycarbonates appear to be more stable toward irradiation than PLA and PLGA. γ-Sterilization of a set of aromatic polycarbonates did not or only slightly affected the surface composition and/or the mechanical properties. Nor was the degradation pattern changed by the sterilization treatment [281]. β-Irradiation of an aliphatic polycarbonate, PTMC, caused a moderate decrease (10%) of Mn [82]. Aseptic processing may be the only useful alternative to achieve sterility of drug delivery systems containing labile polymers and/or labile therapeutic agents.

5.2
Storage

Upon storage in a humid atmosphere, a polymer will absorb water to some extent depending on its chemistry and morphology. Absorbed moisture can react with the biodegradable linkages of the polymer chain, causing an onset of degradation and hence a change in the physicochemical properties [20, 80]. This will, in turn, affect the performance in vivo. The selection of proper storage conditions may thus be of crucial importance to the shelf life of a polymer [190]. For polymers susceptible to hydrolysis, storage under anhydrous conditions (e.g., vacuum, Ar(g), or organic solution) may been required.

For instance, heavy degradation of poly(orthoesters) was observed upon prolonged storage in dry environment [83,177]. In conclusion, POEs have limited shelf life at ambient temperature and special care must be taken to maintain the physical properties of POE during storage. An investigation of the storage stability of PAA, PTMC, and blends thereof after storage in air at 5 °C and 22 °C revealed changes of physicochemical properties, preferably of the polyanhydride component [82]. Compositional changes and a reduction of molecular weight indicated the hydrolytic chain cleavage of PAA during storage, while PTMC appeared stable. Samples stored in Ar(g) were well preserved over time and showed only insignificant changes in composition, Tm, Tg, or molecular weight. Analogous storage-induced hydrolytic breakdown has been reported for other polyanhydrides during storage [75]. This demonstrates the necessity of storing delicate polymers, such as PAA, under anhydrous conditions. A molecular weight decrease has, however, been reported for aliphatic polyanhydrides stored even under anhydrous conditions (e.g., vacuum or organic solution) [319]. This suggests that hydrolysis is not the sole mechanism responsible for breakdown during polyanhydride storage. A self-depolymerization process has been proposed, schematically illustrated in Fig. 16. This anhydride interchange may take place inter- and/or intramolecularly and it results in the formation of oligomeric rings [319].

The purity of a drug delivery device is also an important factor to the storage stability of degradable polymers. The presence of oligomers, residual monomer, or remaining polymerization catalysts or solvents may impair the storage stability, catalyzing moisture absorption or degradation [12, 80, 320]. In polymers,

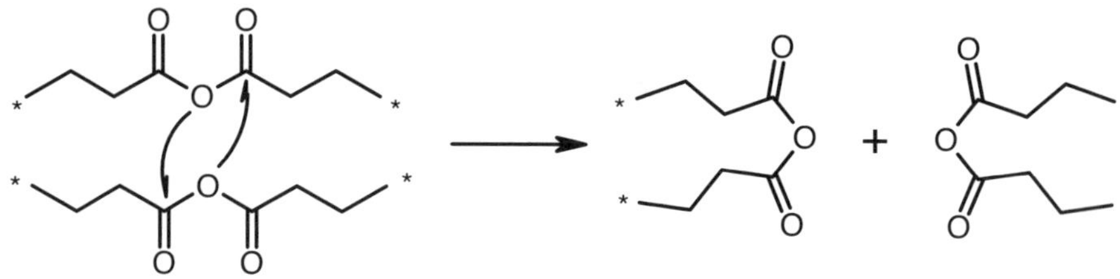

Fig. 16. Schematic illustration of transesterification of polyanhydrides

moisture absorption and diffusivity are strongly influenced by the number and accessibility of polar groups. The incorporation of drug may also affect the storage stability of a polymer matrix [80]. The relative strength of the water-polymer bonds and the degree of crystallization of the polymer matrix are other important factors. High molecular weight PLLA is reported to be stable to atmospheric moisture if it is free from contaminants, since its crystalline regions present an effective barrier to vapor diffusion through the network [170]. Aging studies of PLLA and PDLLA microspheres have shown that storage induces crystallization [68, 190]. The fraction of crystalline material increased with storage time or at increasing relative humidity. This is in agreement with a storage stability study of L-LA:DXO copolymers and blends showing that physicochemical properties were considerably more affected in an atmosphere of high RH than under desiccated conditions [130]. To maintain absolute physicochemical integrity of degradable polymeric drug delivery vehicles, storage in an inert atmosphere is recommended.

6
Conclusions

Conventional drug administration technologies display poor controllability, and lead to high plasma concentrations and short duration times, which frequently lead to adverse effects. Controlled release technology aims at predictable and reproducible delivery of an active substance into a specific environment over an extended period of time, yielding optimal response and prolonged efficiency, and thus offering considerable improvement of many treatments. A powerful approach to controlled drug delivery is the incorporation of the drug into a biodegradable polymeric matrix, which distributes the active substance in a controlled and sustained fashion as the polymer erodes. This review describes degradable polymers commonly used in controlled drug delivery as well as the design of microspheres, applications, biocompatibility and stability of these polymers for controlled and sustained drug delivery.

References

1. Wichterle O, Lim D (1960) Nature 185:117
2. Folkman J, Long DM, Rosenbaum R (1966) Science 154:148
3. Kulkarni RK, Pani KC, Neuman C, Leonard F (1966) Arch Surg 93:839
4. Schmitt EE, Polistina RA (1969) US Patent 3,463,158
5. Schneider AK (1972) US Patent 3,636,956
6. Frazza EJ, Schmitt EE (1971) J Biomed Mater Res Symp 1:43
7. Wasserman D, Versfelt CC (1974) US Patent 3,839,297
8. Benicewicz BC, Hopper PK (1990) J Bioact Compat Polym 5:453
9. Brady JM, Cutright DE, Miller RA, Battistone GC, Hunsuck EE (1973) J Biomed Mater Res 7:155
10. Bos RRM, Rozema FR, Boering G, Nijenhuis AJ, Pennings AJ, Verwey AB, Nieuwenhuis P, Jansen HWB (1991) Biomaterials 12:32
11. Vainionpää S, Rokkanen P, Törmälä P (1989) Prog Polym Sci 14:679

12. Leenslag JW, Pennings AJ, Bos RRM, Rozema FR, Boering G (1987) Biomaterials 8:311
13. Vert M (1989) Angew Makromol Chem 166/167:155
14. Gogolewski S, Pennings AJ (1983) Makromol Chem Rapid Commun 4:675
15. Kronenthal RL (1975) Biodegradable polymers in medicine and surgery. In: Kronenthal RL, Oser Z, Martin E (eds), Polymers in medicine and surgery. Plenum Press, New York, p 119
16. Yolles S, Eldridge JE, Woodland JHR (1971) Polym News 1:9
17. Yolles S (1975) Polym Sci Technol 8:245
18. Higuchi T (1971) US Patent 3,625,214
19. Wise DL, McCormick GJ, Willet GP (1976) Life Sciences 19:867
20. Schindler A, Jeffcoat R, Kimmel GL, Pitt CG, Wall ME, Zweidinger R (1977) Cont Topics Polym Sci 2:251
21. Beck LR, Cowsar DR, Lewis DH, Gibson JW, Flowers CE (1979) Am J Obstet Gynecol 135:419
22. Michaels AS (1976) US Patent 3,962,414
23. Heller J, Baker RW, Gale RM, Rodin JO (1978) J Appl Polym Sci 22:1991
24. Heller J (1984) Crit Rev Ther Drug Carrier Syst 1:39
25. Baker RW, Lonsdale HK (1975) Chem Tech 5:668
26. Ottenbrite RM (1990) Controlled release technology. In: Kroschwitz JI (ed), Encyclopedia of Polymer Science and Engineering, Volume Suppl Vol. Wiley, New York, p 164
27. Peppas NA (1987) Hydrogels in Medicine and Pharmacy, Vol II Polymers. CRC Press, Boca Raton
28. Chasin M, Langer R (1990) Biodegradable polymers as drug delivery systems. Marcel Dekker, New York
29. World market research centre (2000) Business briefing: PharmaTech
30. Davis SS, Illum L, Stolnik S (1996) Curr Opinion Coll Int Sci 1:660
31. Okada H, Ogawa Y, Yashiki T (1987) US Patent 4,652,441
32. Okada H, Yamamoto M, Heya T, Inoue Y, Kamei S, Ogawa Y, Toguchi H (1994) J Control Rel 28:121
33. Göpferich A (1996) Biomaterials 17:103
34. Albertsson A-C, Karlsson S (1994) Chemistry and biochemistry of polymer biodegradation. In: Griffin GJL (ed), Chemistry and Technology of Biodegradable Polymers. Blackie, Glasgow, chap 2, p 7
35. Lenz RW (1993) Adv Polym Sci 107:1
36. Albertsson A-C, Karlsson S (1995) Acta Polym 46:114
37. Amass W, Amass A, Tighe B (1998) Polym Int 47:89
38. Li S, Vert M (1995) Biodegradation of aliphatic polyesters. In: Scott G, Gilead D (eds), Degradable polymers. Chapman & Hall, London, chap 4, p 43
39. Kopecek J, Ulbrich K (1983) Prog Polym Sci 9:1
40. Matlaga BF, Yasenchak LP, Salthouse TN (1976) J Biomed Mater Res 10:391
41. Makino K, Ohshima H, Kondo T (1986) J Microencapsulation 3:203
42. Tamada JA, Langer R (1993) Proc Natl Acad Sci 90:552
43. Göpferich A (1997) Macromolecules 30:2598
44. Siegel RA, Pitt CG (1995) J Control Rel 33:173
45. Medlicott NJ, Tucker IG (1999) Adv Drug Delivery Rev 38:139
46. Juliano RL (1978) Can J Physiol Pharmacol 56:683
47. Fan LT, Singh SK (1989) Controlled release, a quantitative treatment. Springer, Berlin, Heidelberg, New York
48. Ranade VV (1991) J Clin Pharmacol 3:401
49. Chandrasekaran SK, Shaw JE (1977) Con Top in Polym Sci 2:291
50. Singh P, Maibach HI (1994) Crit Rev Ther Drug 11:161
51. Gupta SK, Kumar S, Bolton S, Behl CR, Malick AW (1994) J Control Rel 30:253
52. Vanbever R, Lecouturier N, Preat V (1994) Pharm Res 11:1657
53. Ringsdorf H (1975) J Polym Sci 51:135

54. Domb A, Amselem S, Shah J, Maniar M (1992) Polym Adv Technol 3:279
55. Duncan R, Dimitrijevic S, Evagorou EG (1996) STP Pharm Sci 6:237
56. Ouchi T, Ohya Y (1995) Prog Polym Sci 20:211
57. Oku N (1991) Liposomes. In: Dunn RL, Ottenbrite RM (eds), Polymeric drugs and drug delivery systems. ACS Symp Series 469, chap 3, p 24
58. Dunn RL (1991) Polymeric matrices. In: Dunn RL, Ottenbrite RM (eds), Polymeric drugs and drug delivery systems. ACS Symp Series 469, chap 2, p 11
59. Wood DA (1980) Int J Pharm 7:1
60. Langer R, Peppas NA (1981) Biomaterials 2:201
61. Holland SJ, Tighe BJ, Gould PL (1986) J Control Rel 4:155
62. Brannon-Peppas L (1995) Int J Pharm 116:1
63. Allémann E, Leroux J-C, Gurny R (1998) Adv Drug Delivery Rev 34:171
64. Wang YM, Sato H, Horikoshi I (1997) J Control Rel 49:157
65. Beck LR, Flowers CF, Cowsar DR, Tanquary AC (1988) US Patent 4,732,763
66. O'Hagan DT (1998) Adv Drug Del Rev 34:305
67. Sturesson C, Artursson P, Ghaderi R, Johansen K, Mirazimi A, Uhnoo I, Svensson L, Albertsson A-C, Carlfors J (1999) J Control Rel 59:377
68. Jalil R, Nixon JR (1990) J Microencapsulation 7:297
69. Schmitt EA, Flanagan DR, Linhardt RJ (1994) Macromolecules 27:743
70. Higuchi T (1961) J Pharm Sci 50:874
71. Ritger PL, Peppas NA (1987) J Control Rel 5:23
72. Wada R, Hyon S-H, Ikada Y.(1995) J Control Rel 37:151
73. Batycky RP, Hanes J, Langer R, Edwards DA (1997) J Pharm Sci 86:1464
74. Yoshida R, Sakai K, Okano T, Sakurai Y (1991) Polym J 23:1111
75. Leong KW, Brott BC, Langer R (1985) J Biomed Mater Res 19:941
76. Li Y-X, Feng X-D (1990) Makromol Chem Makromol Symp 33:253
77. Heller J (1980) Biomaterials 1:51
78. Szycher M (1992) Szycher's Dictionary of Biomaterials and Medical Devices. Technomic Publishing Company, Lancaster
79. Williams DF (1989) J Biomed Eng 11:185
80. Merkli A, Heller J, Tabatabay C, Gurny R (1996) Biomaterials 17:897
81. Anderson JM, Langone JJ (1999) J Control Rel 57:107
82. Edlund U, Albertsson A-C, Singh SK, Fogelberg I, Lundgren BO (2000) Biomaterials 21:945
83. Zignani M, Bernatchez SF, Le Minh T, Tabatabay C, Anderson JM, Gurny R (1998) J Biomed Mater Res 39:277
84. Sakellariou P, Rowe RC (1995) Prog Polym Sci 20:889
85. Pereswetoff-Morath L (1998) Adv Drug Del Rev 29:185
86. Rege PR, Shukla DJ, Block LH (1999) Int J Pharm 181:49
87. Lorenzo-Lamosa ML, Remunan-Lopez C, Vila-Jato JL, Alonso MJ (1998) J Control Rel 52:109
88. Fujioka K, Maeda M, Hojo T, Sano A (1998) Adv Drug Delivery Rev 31:247
89. Tabata Y, Ikada Y (1998) Adv Drug Delivery Rev 31:287
90. Lewis DH (1990) Controlled release of bioactive agents from lactide/glycolide polymers. In: Chasin M, Langer R (eds), Biodegradable polymers as drug delivery systems. Marcel Dekker, New York, chap 1, p 1
91. Grandfils C, Flandroy P, Nihant N, Barbette S, Jérome R, Teyssié Ph, Thibaut A (1992) J Biomed Mater Res 26:467
92. Kricheldorf HR, Kreiser-Saunders I (1996) Macromol Symp 103:85
93. Zhang X, Wyss UP, Pichora D, Goosen MFA (1994) J Bioact Compat Polym 9:80
94. Polard E, Le Corre P, Chevanne F, Le Verge R (1996) Int J Pharm 134:37
95. Vert M, Li SM, Spenlehauer, G, Guerin P (1992) J Mater Sci Mater Med 3:432
96. Heya T, Okada H, Ogawa Y, Toguchi H (1991) Int J Pharm 72:199
97. Chiu LK, Chiu WJ, Cheng Y-L (1995) Int J Pharm 126:169

98. Cohen S, Chen L, Apte RN (1995) Reactive Polym 25:177
99. Crotts G, Park TG (1998) J Microencapsulation 15:699
100. Cheng Y-H, Illum L, Davis SS (1998) J Control Rel 55:203
101. Coombes AGA, Yeh M-K, Lavelle EC, Davis SS (1998) J Control Rel 52:311
102. Cao X, Shoichet MS (1999) Biomaterials 20:329
103. Castelli F, Giunchedi P, LaCamera O, Conte U (2000) Drug Delivery 7:1
104. Chandrashekar G, Udupa N (1996) J Pharm Pharmacol 48:669
105. Dubernet C, Benoit JP, Couarraze G, Duchêne D (1987) Int J Pharm 35:145
106. Pitt CG (1990) Poly-ε-caprolactone and its copolymers. In: Chasin M, Langer R (eds), Biodegradable polymers as drug delivery systems. Marcel Dekker, New York, chap 3, p 71
107. Pouton CW, Akhtar S (1996) Adv Drug Del Syst 18:133
108. Martin MA, Miguens FC, Rieumont J, Sanchez R (2000) Coll Surf B 17:111
109. Nobes GAR, Marchessault RH, Maysinger D (1998) Drug Delivery 5:167
110. Holland SJ, Jolly AM, Yasin M, Tighe BJ (1987) Biomaterials 8:289
111. Chasin M, Domb A, Ron E, Mathiowitz E, Langer R, Leong K, Laurencin C, Brem H, Grossman S (1990) Polyanhydrides as drug delivery systems. In: Chasin M, Langer R (eds), Biodegradable Polymers as Drug Delivery Systems. Marcel Dekker, New York, chap 2, p 43
112. Leong KW, D'Amore P, Marletta M, Langer R (1986) J Biomed Mater Res 20:51
113. Mathiowitz E, Amato C, Dor Ph, Langer R (1990) Polymer 31:547
114. Maniar M, Domb A, Haffer A, Shah J (1994) J Control Rel 30:233
115. Wu MP, Tamada JA, Brem H, Langer R (1994) J Biomed Mater Res 28:387
116. Park E-S, Maniar M, Shah JC (1997) J Control Rel 48:67
117. Mathiowitz E, Jacob JS, Jong YS, Carino GP, Chickering DE, Chaturvedi P, Santos CA, Vijayaraghavan K, Montgomery S, Basset M, Morrell C (1997) Nature 386:410
118. Chickering D, Jacob J, Mathiowitz E (1996) Biotech Bioeng 52:96
119. Leach KJ, Takahashi S, Mathiowitz E (1998) Biomaterials 19:1981
120. Zhu KJ, Hendren RW, Jensen K, Pitt CG (1991) Macromolecules 24:1736
121. Edlund U, Albertsson A-C (1999) J Appl Polym Sci 72:227
122. Heller J (1985) J Control Rel 2:167
123. Sparer RV, Shih C, Ringeisen CD, Himmelstein KJ (1984) J Control Rel 1:23
124. Maa YF, Heller J (1990) J Control Rel 14:21
125. Heller J, Chang AC, Rodd G, Grodsky GM (1990) J Control Rel 13:295
126. Heller J, Ng SY, Fritzinger BK, Roskos KV (1990) Biomaterials 11:235
127. Roskos KV, Fritzinger BK, Rao SS, Armitage GC, Heller J (1995) Biomaterials 16:313
128. Zignani M, Merkli A, Sintzel MB, Bernatchez SF, Kloeti W, Heller J, Tabatabay C, Gurny R (1997) J Control Rel 48:115
129. Edlund U, Albertsson A-C (1999) J Polym Sci A Polym Chem 37:1877
130. Edlund U, Albertsson A-C (2000) J Polym Sci A Polym Chem 38:786
131. Edlund U, Albertsson A-C (2000) J Bioact Compat Polym 15:214
132. Cook TJ, Amidon GL, Yang VC (1997) Int J Pharm 159:197
133. Kim KS, Kim TK, Graham NB (1999) Polym J 31:813
134. Zhao X, Harris JM (1998) J Pharm Sci 87:1450
135. La SB, Nagasaki Y, Kataoka K (1997) ACS Symp Ser 680:99
136. Lu S, Anseth KS (2000) Macromolecules 33:2509
137. Bae YH, Huh KM, Kim Y, Park K-H (2000) J Control Rel 64:3
138. Schacht E, Vandorpe J, Dejardin S, Lemmouchi Y, Seymour L (1996) Biotechnol Bioeng 52:102
139. Ibim SM, Ambrosio AA, Larrier D, Allcock HR, Laurencin CT (1996) J Control Rel 40:31
140. Andrianov AK, Payne LG (1998) Adv Drug Delivery Rev 31:185
141. Domb AJ (1993) J Polym Sci A Polym Chem 31:1973
142. Carothers WH, Arvin GA (1929) J Am Chem Soc 51:2560

143. Pitt CG, Gratzl MM, Kimmel GL, Surles J, Schindler A (1981) Biomaterials 2:215
144. Grijpma DW, Pennings AJ (1994) Macromol Chem Phys 195:1633
145. Tsuji H, Ikada Y (2000) Polym Degr Stab 67:179
146. Williams DF (1981) Eng Med 10:5
147. Pitt CG, Gu Z (1987) J Control Rel 4:283
148. Pistner H, Gutwald R., Ordung R, Reuther J, Mühling J (1993) Biomaterials 14:671
149. Lundberg RD, Cox EF (1969) Lactones. In: Frisch KC, Reegen SL (eds), Ring-opening polymerization. Marcel Dekker, New York, chap 6, p 247
150. Löfgren A, Albertsson A-C, Dubois Ph, Jérôme R (1995) J Macromol Sci Rev Macromol Chem Phys C 35:379
151. Albertsson A-C, Ljungquist O (1986) J Macromol Sci Chem A 23:393
152. Albertsson A-C, Ljungquist O (1988) Acta Polym 39:95
153. Mathisen T, Masus K, Albertsson A-C (1989) Macromolecules 22:3842
154. Löfgren A, Albertsson A-C (1995) J Macromol Sci Pure Appl Chem A 32:41
155. Gruvegård M, Lindberg T, Albertsson A-C (1998) J Macromol Sci Pure Appl Chem A 35:885
156. Stridsberg K, Gruvegård M, Albertsson A-C (1998) Macromol Symp 130:367
157. Stridsberg K, Ryner M, Albertsson A-C (2000) Macromolecules 33:2862
158. Brode GL, Koleske JV (1972) J Macromol Sci Chem A 6:1109
159. Dubois Ph, Ropson N, Jérôme R, Teyssié Ph (1996) Macromolecules 29:1965
160. Jedlinski Z, Kurcok P, Kowalczuk M, Matuszowicz A, Dubois P, Jerome R, Kricheldorf HR (1995) Macromolecules 28:7276
161. Kreiser-Saunders I, Kricheldorf HR (1998) Macromol Chem Phys 199:1081
162. Kricheldorf HR, Kreiser-Saunders I, Boettcher C (1995) Polymer 36:1253
163. Kricheldorf HR, Eggerstedt S (1998) Macromol Chem Phys 199:283
164. Kowalski A, Duda A, Penczek S (1998) Macromol Rapid Commun 19:567
165. Stevels WM, Ankoné MJK, Dijkstra PJ, Feijen J (1996) Macromolecules 29:6132
166. Deng XM, Yuan ML, Xiong CD, Li XH (1999) J Appl Polym Sci 71:1941
167. Simic V, Girardon V, Spassky N, Hubert-Pfalzgraf LG, Duda A (1998) Polym Degr Stab 59:227
168. Chamberlain BM, Sun Y, Hagadorn JR, Hemmesch EW, Young VG, Pink M, Hillmyer MA, Tolman WB (1999) Macromolecules 32:2400
169. Mecerreyes D, Jérôme R (1999) Macromol Chem Phys 200:2581
170. Vert M, Schwarch G Coudane J (1995) J Macromol Sci Pure Appl Chem A 32:787
171. Kobayashi S, Uyama H, Namekawa S (1998) Polym Degr Stab 59:195
172. Dong H, Cao S-G, Li Z-Q, Han S-P, You D-L, Shen J-C (1999) J Polym Sci A Polym Chem 37:1265
173. Henderson LA, Svirkin YY, Gross RA, Kaplan DL, Swift G (1996) Macromolecules 29:7759
174. Matsumura S, Mabuchi K, Toshima K (1998) Macromol Symp 130:285
175. Kulkarni RK, Moore EG, Hegyeli AF, Leonard F (1971) J Biomed Mater Res 5:169
176. Thérin M, Christel P, Li S, Garreau H, Vert M (1992) Biomaterials 13:594
177. Engelberg I, Kohn J (1991) Biomaterials 12:292
178. Lunt J (1998) Polym Degr Stab 59:145
179. Gilding DK, Reed AM (1979) Polymer 20:1459
180. Feng X-D, Jia Y (1997) Macromol Symp 118:625
181. Spenlehauer G, Vert M, Benoît J-P, Chabot F, Veillard M (1988) J Control Rel 7:217
182. Duda A, Biela T, Libiszowski J, Penczek S, Dubois P, Mecerreyes D, Jérôme R (1998) Polym Degr Stab 59:215
183. Li SM, Rashkov I, Espartero JL, Manolova N, Vert M (1996) Macromolecules 29:57
184. Yasugi K, Nagasaki Y, Kato M, Kataoka K (1999) J Control Rel 62:89
185. Metters AT, Anseth KS, Bowman CN (2000) Polymer 41:3993
186. Kim J-H, Emoto K, Iijima M, Nagasaki Y, Aoyagi T, Okano T, Sakurai Y, Kataoka K (1999) Polym Adv Technol 10:647

187. Tsuji H, Ikada Y (1998) J Appl Polym Sci 67:405
188. Cha Y, Pitt CG (1988) J Control Rel 7:69
189. Seki T, Kawaguchi T, Endoh H, Ishikawa K, Juni K, Nakano M (1990) J Pharm Sci 79:985
190. Aso Y, Yoshioka S, Terao T (1993) Int J Pharm 93:153
191. Leo E, Forni F, Bernabei MT (2000) Int J Pharm 196:1
192. Gander B, Johansen P, Nam-Trân H, Merkle HP (1996) Int J Pharm 129:51
193. Luo D, Woodrow-Mumford K, Belcheva N, Saltzman WM (1999) Pharm Res 16:1300
194. Reed AM, Gilding DK (1981) Polymer 22:494
195. Tuncay M, Calis S, Kas HS, Ercan MT, Peksoy I, Hincal AA (2000) Int J Pharm 195:179
196. Yang J, Cleland JL (1997) J Pharm Sci 86:908
197. Park TG, Lee HY, Nam YS (1998) J Control Rel 55:181
198. Kim HK, Park TG (1999) Biotech Bioeng 65:659
199. Péan J-M, Venier-Julienne M-C, Boury F, Menei P, Denizot B, Benoit J-P (1998) J Control Rel 56:175
200. Maulding HV (1987) J Control Rel 6:167
201. Sah H, Toddywala R, Chien YW (1995) J Control Rel 35:137
202. Blanco-Prieto MJ, Besseghir K, Zerbe O, Andris D, Orsolino P, Heimgartner F, Merkle HP, Gander B (2000) J Control Rel 67:19
203. Metha RC, Thanoo BC, DeLuca PP (1996) J Control Rel 41. 249
204. Thompson WW, Anderson DB, Heiman ML (1997) J Control Rel 43:9
205. Walter E, Moelling K, Pavlovic J, Merkle HP (1999) J Control Rel 61:361
206. Tinsley-Bown AM, Fretwell R, Dowsett AB, Davis SL, Farrar GH (2000) J Control Rel 66:229
207. Wang D, Robinson DR, Kwon GS, Samuel J (1999) J Control Rel 57:9
208. Ertl B, Plazer P, Wirth M, Gabor F (1999) J Control Rel 61. 305
209. Mohr D, Wolff M, Kissel T (1999) J Control Rel 61:203
210. Birnbaum DT, Kosmala JD, Henthorn DB, Brannon-Peppas L (2000) J Control Rel 65:375
211. Mogi T, Ohtake N, Yoshida M, Chimura R, Kamaga Y, Ando S, Tsukamoto T, Nakajima T, Uenodan H, Otsuka M, Matsuda Y, Ohshima H, Makino K (2000) Colloid Surface B 17:153
212. Philips M, Gresser JD (1984) J Pharm Sci 73:1718
213. Stoller NH, Johnson LR, Trapnell S, Harrold CQ, Garett S (1998) J Periodontol 69:1085
214. Benedict CV, Cameron JA, Huang SJ (1983) J Appl Polym Sci 28:335
215. Eldsäter C, Erlandsson B, Renstad R, Albertsson A-C, Karlsson S (2000) Poylmer 41:1297
216. Rutkowska M, Dereszewska A, Jastrzebska M, Janik H (1998) Macromol Symp 130:199
217. Chen DR, Bei JZ, Wang SG (2000) Polym Degr Stab 67:455
218. Lemmouchi Y, Schacht E, Kageruka P, De Deken R, Diarra B, Diall O, Geerts S (1998) Biomaterials 19:1827
219. Lostocco MR, Murphy CA, Cameron JA, Huang SJ (1998) Polym Degr Stab 59:303
220. Tian D, Dubios Ph, Jerome R (1996) Polymer 37:3983
221. Eastmond GC (1999) Adv Polym Sci 149:59
222. Perez MH, Zinutti C, Lamprecht A, Ubrich N, Astier A, Hoffman M, Bodmeier R, Maincent P (2000) J Control Rel 65:429
223. Vandamme TF, Mukendi JFN (1996) Int J Pharm 145:77
224. Jameela SR, Suma N, Jayakrishnan A (1997) J Biomater Sci Polym Ed 8:466
225. Goodwin CJ, Braden M, Downes S, Marshall NJ (1998) J Biomed Mater Res 40:204
226. Buntner B, Nowak M, Kasperczyk J, Ryba M, Grieb P, Walski M, Dobrzyński P, Bero M (1998) J Control Rel 56:159
227. Allen C, Han J, Yu Y, Maysinger D, Eisenberg A (2000) J Control Rel 63:275
228. Merrick JM, Doudoroff M (1961) Nature 189:890
229. Doi Y, Kunioka M, Nakamura Y, Soga K (1986) Macromolecules 19:1274
230. Sharma R, Ray AR (1995) J Macromol Sci Rev Macromol Chem Phys C 35:327
231. Holmes PA (1985) Phys Technol 16:32

232. Kawaguchi Y, Doi Y (1992) Macromolecules 25:2324
233. Doi Y (1995) Macromol Symp 98:585
234. Doi Y (2000) Abstr Pap – IUPAC Macro 2000, July 2000, Warsaw, Poland 1:8
235. Gross RA, DeMello C, Lenz RW, Brandl H, Fuller RC (1989) Macromolecules 22:1106
236. Eldsäter C, Albertsson A-C, Karlsson S (1997) Acta Polymer 48:478
237. Miller ND, Williams DF Biomaterials (1987) 8:129
238. Holland SJ, Yasin M, Tighe BJ (1990) Biomaterials 11:206
239. Yasin M, Tighe BJ (1992) Biomaterials 13:9
240. Atkins TW (1997) Biomaterials 18:173
241. Atkins TW, Peacock SJ (1996) J Microencapsulation 13:709
242. Gangrade N, Price JC (1991) J Microencapsulation 8:185
243. Sendil D, Gürsel I, Wise DL, Hasirci V (1999) J Control Rel 59:207
244. Juni K, Nakano M, Kubota M (1986) J Control Rel 4:25
245. Abe H, Doi Y, Yamamoto Y (1992) J Macromol Sci Pure Appl Chem 29:229
246. Kawaguchi T, Tsugane A, Higashide K, Endoh H, Hasegawa T, Kanno H, Seki T, Juni K, Fukushima S, Nakano M (1992) J Pharm Sci 81:508
247. Bucher JE, Slade WC (1909) J Am Chem Soc 31:1319
248. Hill JW (1930) J Am Chem Soc 52:4110
249. Mathiowitz E, Langer R (1987) J Control Rel 5:13
250. Tamargo RJ, Epstein JI, Reinhard CS, Chasin M, Brem H (1989) J Biomed Mater Res 23:253
251. Albertsson A-C, Lundmark S (1988) J Macromol Sci Chem A 25:247
252. Albertsson A-C, Lundmark S (1990) Br Polym J 23:205
253. Domb AJ, Amselem S, Shah J, Maniar M (1993) Adv Polym Sci 107:94
254. Domb AJ, Langer R (1987) J Polym Sci A Polym Chem 25:3373
255. Albertsson A-C, Lundmark S (1990) J Macromol Sci Chem A 27:397
256. Lundmark S, Sjöling M, Albertsson A-C (1991) J Macromol Sci Chem A 28:15
257. Albertsson A-C, Eklund M (1995) J Appl Polym Sci 57:87
258. Shieh L, Tamada J, Chen I, Pang J, Domb A, Langer R (1994) J Biomed Mater Res 28:1465
259. Mathiowitz E, Ron E, Mathiowitz G, Amato C, Langer R (1990) Macromolecules 23:3212
260. Göpferich A, Langer R (1993) J Polym Sci A Polym Chem 31:2445
261. Shuai X, Tan H, Jedlinski Z (1997) Polym Bull 39:21
262. Teomim D, Nyska A, Domb AJ (1999) J Biomed Mater Res 45, 258
263. Gouin S, Zhu XX, Lehnert S (2000) Macromolecules 33:5379
264. Uhrich KE, Gupta A, Thomas TT, Laurencin CT, Langer R (1995) Macromolecules 28:2184
265. Chiba M, Hanes J, Langer R (1997) Biomaterials 18:893
266. Jiang HL, Zhu KJ (1999) Polym Int 48:47
267. Albertsson A-C, Carlfors J, Sturesson C (1996) J Appl Polym Sci 62:695
268. Dang W, Daviau T, Brem H (1996) Pharm Res 13:683
269. Mayberg MR, Gross AS, Mathiowitz E, Langer R (1992) Polym Adv Technol 3:331
270. Teomim D, Fishbien I, Golomb G, Orloff L, Mayberg M, Domb AJ (1999) J Control Rel 60:129
271. Stephens D, Li L, Robinson D, Chen S, Chang H-C, Liu RM, Tian Y, Ginsburg EJ, Gao X, Stultz T (2000) J Control Rel 63:305
272. Erdmann L, Uhrich KE (2000) Biomaterials 21:1941
273. Carothers WH, van Natta FJ (1930) J Am Chem Soc 52:314
274. Kricheldorf HR, Jenssen J (1989) J Macromol Sci Chem A 26:631
275. Kricheldorf HR, Dunsing R, Serra I, Albet A (1987) Makromol Chem 188:2453
276. Matsumura S, Tsukada K, Toshima K (1997) Macromolecules 30:3122
277. Bisht KS, Svirkin YY, Henderson LA, Gross RA, Kaplan DL, Swift G (1997) Macromolecules 30:7735

278. Ranucci E, Liu Y, Lindblad MS, Albertsson A-C (2000) Macromol Rapid Comm 21:680
279. Albertsson A-C, Sjöling M (1992) J Macromol Sci Pure Appl Chem A 29:43
280. Albertsson A-C, Eklund M (1994) J Polym Sci A Polym Chem 32:265
281. Hooper KA, Cox JD, Kohn J (1997) J Appl Polym Sci 63:1499
282. McNeill IC, Rincon A (1989) Polym Degr Stab 24:59
283. Buchholz B (1993) J Mater Sci Mater Med 4:381
284. Joziasse CA, Grablowitz H, Pennings AJ (2000) Macromol Chem Phys 201:107
285. Capozza RC (1978) US Patent 4,066,747
286. Choi NS, Heller J (1978) US Patent 4,079,038
287. Choi NS, Heller J (1978) US Patent 4,093,709
288. Choi NS, Heller J (1979) US Patent 4,138,344
289. Heller J (1990) Drugs Pharm Sci 45:121
290. Heller J, Helwing RF, Penhale DW (1981) US Patent 4,304,767
291. Heller J, Roskos KV, Duncan R (1993) Makromol Chem Macromol Symp 70/71:163
292. Merkli A, Heller J, Tabatabay C, Gurny R (1995) J Control Rel 33:415
293. Albertsson A-C, Palmgren R (1993) J Macromol Sci Pure Appl Chem A 30:919
294. Ray JA, Doddi N, Regula D, Williams JA, Melveger A (1981) Surg Gynecol 153:497
295. Nishida H, Yamashita M, Nagashima M, Endo T, Tokiwa Y (2000) J Polym Sci A Polym Chem 38:1560
296. Kafrawy A, Mattei FV, Shalaby SW (1984) US Patent 4,470,416
297. Kafrawy A, Shalaby SW (1986) J Bioact Compat Polym 1:431
298. Mathisen T, Albertsson A-C (1989) Macromolecules 22:3838
299. Löfgren A, Albertsson A-C, Dubois Ph, Jérôme R, Teyssié Ph (1994) Macromolecules 27:5556
300. Stridsberg K, Albertsson A-C (1999) J Polym Sci A Polym Chem 37:3407
301. Palmgren R, Karlsson S, Albertsson A-C (1997) J Polym Sci A Polym Chem 35:1635
302. Karlsson S, Hakkarainen M, Albertsson A-C (1994) J Chromatogr A 688:251
303. Stridsberg K, Albertsson A-C (2000) J Polym Sci A Polym Chem 38:1774
304. Flinn JE, Nack H (1967) Chem Eng (New York) 75:171
305. Watts PJ, Davies MC, Melia CD (1990) Crit Rev Ther Drug Carrier Syst 7:235
306. Jain R, Shah NH, Malick AW, Rhodes CT (1998) Drug Dev Ind Pharm 24:703
307. Arshady R (1991) J Control Rel 17:1
308. Bakan JA (1975) Polym Sci Technol 8:213
309. Tice TR, Lewis DH (1983) US Patent 4,389,330
310. Viswanathan NB, Thomas PA, Pandit JK, Kulkarni MG, Mashelkar RA (1999) J Control Rel 58:9
311. Nuwayser ES, Nucefora WA (1986) US Patent 4,623,588
312. Booth AF (2000) Sterilization validation and routine operation handbook. Technomic Publishing Company, Lancaster
313. Nair PD (1995) J Biomater Appl 10:121
314. Sintzel MB, Merkli, A, Tabatabay C, Gurny R (1997) Drug Dev Ind Pharm 23:857
315. Bruck SD, Mueller EP (1988) J Biomed Mater Res (Appl Biomater) 22:133
316. Burg KJL, Shalaby SW (1996) ACS Symp Ser Irradiation of Polymers 620:240
317. Birkinshaw C, Buggy M, Henn GG (1992) Polym Degr Stab 38:249
318. Mäder K, Domb A, Swartz HM (1996) Appl Radiat Isot 47:1669
319. Domb AJ, Langer R (1989) Macromolecules 22:2117
320. Hyon S-H, Jamshidi K, Ikada Y (1998) Polym Int 46:196

Received: January 2001

Aliphatic Polyesters: Abiotic and Biotic Degradation and Degradation Products

Minna Hakkarainen

Department of Polymer Technology, Royal Institute of Technology (KTH), 10044 Stockholm, Sweden
e-mail: minna@polymer.kth.se

Abstract. This paper reviews the degradation behavior of aliphatic polyesters of current interest, including polylactide, polycaprolactone, poly(3-hydroxybutyrate) and their copolymers. Special focus is given to degradation products formed in different abiotic and biotic environments. The influence of processing and processing additives on the properties and degradation behavior is also briefly discussed.

Keywords. Polylactide, Polycaprolactone, Poly(3-hydroxybutyrate), Degradation, Degradation products

1	Introduction 115	
1.1	Polylactide and its Copolymers 115	
1.2	Polycaprolactone and its Copolymers 117	
1.3	Poly(3-hydroxybutyrate) and its Copolymers 117	
2	Degradation of Aliphatic Polyesters 118	
2.1	Degradation of Polylactide 118	
2.2	Degradation of Polycaprolactone 121	
2.3	Degradation of PHB and PHBV 124	
3	Degradation Products of Aliphatic Polyesters 125	
3.1	Degradation Products of PLA and its Copolymers 126	
3.2	Degradation Products of PCL and its Copolymers 129	
3.4	Degradation Products of PHB and its Copolymers 130	
4	The Influence of Processing and Processing Additives 131	
4.1	The Influence of Processing on the Properties of Polylactide 132	
4.2	The influence of Processing on the Properties of Polycaprolactone . 132	
4.3	The Influence of Processing on the Properties of PHB and PHBV . 133	

5 Conclusions 134

References 134

Abbreviations

$CaCO_3$	calcium carbonate
CL	caprolactone
CZE	capillary zone electrophoresis
DSC	differential scanning calorimetry
DXO	1,5-dioxepan-2-one
ER	erucamide
GA	glycolide
GC-MS	gas chromatography-mass spectrometry
3HB	3-hydroxybutyric acid
3HH	3-hydroxyhexanoic acid
6HH	6-hydroxyhexanoic acid
HPLC	high performance liquid chromatography
3HV	3-hydroxyvaleric acid
LA	lactide
LC	liquid chromatography
LDPE	low density polyethylene
LLDPE	linear low density polyethylene
LLE	liquid-liquid extraction
MALDI-TOF	matrix assisted laser desorption ionization-time of flight
MHE	multiple headspace extraction
M_n	number-average molecular weight
M_w	weight-average molecular weight
NMR	nuclear magnetic resonance
PCL	polycaprolactone
PDXO	poly(1,5-dioxepan-2-one)
PET	poly(ethylene terephthalate)
PHA	polyhydroxyalkanoate
PHB	poly(3-hydroxybutyrate)
PHBV	poly(3-hydroxybutyrate-*co*-3-hydroxyvalerate)
PDLLA	poly(D,L-lactide)
PGA	polyglycolide
PLA	polylactide
PLG	poly(lactide-*co*-glycolide)
PLLA	poly(L-lactide)
PP	polypropylene
PS	polystyrene
PTMC	poly(trimethylene carbonate)
PVAl	polyvinyl alcohol
PVC	polyvinyl chloride

SEM	scanning electron microscope
SiO_2	silicon dioxide
SPE	solid phase extraction
SPME	solid phase microextraction
TMC	trimethylene carbonate

1
Introduction

Research in degradable polymers has gained considerable interest in recent years due to the increasingly attractive environmental, biomedical, and agricultural applications. Aliphatic polyesters are one of the most promising materials to be used as, e.g., packaging materials and mulch films to solve the problems related to plastic waste accumulation. The development of biodegradable products such as fishing lines and nets could solve the problems caused by the several hundred thousand tons of non-degradable plastic products discarded into marine environments every year causing death of numerous marine animals. Polylactide (PLA), polycaprolactone (PCL), poly(3-hydroxybutyrate) (PHB), and their copolymers are the most studied aliphatic polyesters for a wide variety of applications. A large range in the properties and degradation behavior is obtained by, e.g., copolymerization and blending.

1.1
Polylactide and its Copolymers

Polylactide is perhaps the most frequently used polyester in biomedical applications due to its many favorable characteristics, e.g., high strength and biocompatibility. Environmental concern has led to escalated interest in using polylactide and other biodegradable polymers as an alternative to traditional commodity plastics. PLA is synthesized either from lactic acid by a condensation reaction or more commonly by ring-opening polymerization of cyclic lactide monomer. The lactide monomer exists in three different forms: two stereoisomers L- and D-lactide (L- and D-LA) and racemic D,L-lactide (*meso*-lactide). The chirality of LA units provides a means to adjust degradation rates, as well as physical and mechanical properties. Poly(L-lactide) (PLLA) is a semicrystalline polymer with a relatively slow hydrolysis rate. Poly(D,L-lactide) (PDLLA) is an amorphous polymer with a much faster hydrolysis rate. T_m and T_g for PLLA are around 215 °C and 55–58 °C, respectively [1]. In addition to combining the different stereoisomers of lactic acid, copolymerization with glycolide (GA) [2, 3] or other monomers like ε-caprolactone (CL) [4–6], 1,5-dioxepan-2-one (DXO) [7, 8], trimethylene carbonate (TMC) [9] is used to design polymers with widely different properties and degradation times. The properties can also be modified by blending high and low molecular weight polylactide [10] or by blending polylactide with other polyesters [11–13]. The chemical structures of several cyclic monomers and the resulting PLA-copolymers is presented in Table 1.

Table 1. Monomers and resulting repeating units of polylactide and some of its copolymers

Monomer	Polymer
lactide	polylactide $H-[O-CH(CH_3)-C(=O)]-OH$
lactide, glycolide	poly(lactide-co-glycolide) $H-[O-CH(CH_3)-C(=O)]-[O-CH_2-C(=O)]-OH$
lactide, ε-caprolactone	poly(lactide-co-caprolactone) $H-[O-CH(CH_3)-C(=O)]-[O-(CH_2)_5-C(=O)]-OH$
lactide, 1,5-dioxepan-2-one (DXO)	poly(lactide-co-1,5-dioxepan-2-one) $H-[O-CH(CH_3)-C(=O)]-[O-(CH_2)_2-O-(CH_2)_2-C(=O)]-OH$
lactide, trimethylene carbonate (TMC)	poly(lactide-co-trimethylene carbonate) $H-[O-CH(CH_3)-C(=O)]-[O-(CH_2)_3-O-C(=O)]-OH$

The physical properties and melt processing of PLA are similar to those of conventional packaging resins. It may thus be used as a commodity resin for general packaging application. In many aspects the basic properties of PLA lie between those of crystal PS and PET [14]. When plasticized by its own monomer lactic acid, PLA becomes increasingly flexible so that products that mimic PVC, LDPE, LLDPE, PP, and PS can be prepared [15]. Possible applications are especially disposable materials such as food packaging, diapers, contaminated hospital waste, which are not suitable for collecting and recycling. PLA polymers are also used in a broad variety of medical applications: bioresorbable surgical sutures, dental implants, bone screws and plates, controlled drug delivery, and so on.

1.2
Polycaprolactone and its Copolymers

Polycaprolactone is a linear polyester, manufactured by ring-opening polymerization of ε-caprolactone. It is a semicrystalline polymer with a degree of crystallinity around 50%. It has a rather low glass transition temperature (T_g=–60 °C) and melting point (T_m=60 °C). The PCL chain is flexible and exhibits high elongation at break and low modulus. Polycaprolactone has been blended with various polymers and fillers, e.g., PVC [16], PET [17], polyvinyl alcohol (PVAl) [18], PE [19], and tropical starch [20]. However, in most cases the mechanical properties or the degradability were reduced compared to the homopolymer. PCL has also been blended and copolymerized with PLA [12, 21], PHB [22, 23], and PDXO [24, 25]. The copolymers of PCL and PDXO were crystalline up to a DXO content of 40% and it was concluded that the DXO units were incorporated into the PCL crystals. The block copolymers of CL and DXO exhibit thermoplastic elastomeric properties [26].

1.3
Poly(3-hydroxybutyrate) and its Copolymers

Poly(3-hydroxybutyrate) is a naturally occurring polyester produced by numerous bacteria in nature as an intracellular reserve of carbon or energy. PHB was first discovered in *Bacillus megaterium* in 1925 by Lemoigne [27]. Since then it has been found in a large number of bacteria. The bacterium *Alcaligenes eutrophus* is one of the most frequently used microorganisms for the biosynthesis of polyhydroxyalkanoates (PHAs) [28–30]. The PHB granules in intact cells are completely amorphous, but they crystallize after extraction [31]. PHB is often compared to polypropylene with regard to its physical properties because they have similar melting points, degree of crystallinity, and glass transition temperatures [32, 33]. In general, PHB is stiffer and more brittle than polypropylene. In addition, PHB exhibits much lower solvent resistance but better natural resistance to ultraviolet radiation than polypropylene [32]. The properties of PHB are often modified by copolymerization with 3-hydroxyvaleric acid (3HV). Poly(3-hydroxybutyrate-*co*-3-hydroxyvalerate) (PHBV) copolymers have been produced in compositions ranging from 0–90% HV content [28]. PHBV finds applications in various packaging materials, such as thin films and paper coatings [34].

2
Degradation of Aliphatic Polyesters

The degradation of aliphatic polyesters proceeds by one or several mechanisms including chemical hydrolysis, microbial, enzymatic and thermal degradation. The degradation proceeds either at the surface (homogeneous) or within the bulk (heterogeneous) and is controlled by a wide variety of compositional and property variables, e.g., matrix morphology, chain orientation, chemical composition, stereochemical structure, sequence distribution, molecular weight and distribution, the presence of residual monomers, oligomers and other low molecular weight products, size and shape of the specimen, and the degradation environment, e.g., presence of moisture, oxygen, microorganisms, enzymes, pH, temperature, and so on. Which degradation mechanism dominates depends on both the structure of the polyester and the environment it is subjected to.

2.1
Degradation of Polylactide

Most of the studies on PLA degradation have concentrated on abiotic hydrolysis [35–37]. The effects of, e.g., residual monomer and other impurities, molecular weight and copolymerization on hydrolysis rate and properties have been studied [3, 37–42]. Impurities, residual monomer [43, 44], and peroxide modification [45] all increase the hydrolysis rate, while copolymerization can either increase (GA-copolymers) or decrease (CL, DXO-copolymers) the hydrolysis rate. Degradation of PLA and its copolymers in clinical applications ranging from absorbable sutures to drug delivery systems and artificial ligaments has also been widely studied [46–48].

The degradation of semicrystalline polyesters in aqueous media occurs in two stages [49, 50]. The first stage starts with water diffusion into the amorphous regions, which are less organized and allow water to penetrate more easily. The second stage starts when most of the amorphous regions are degraded. The hydrolytic attack then proceeds from the edge toward the center of the crystalline domains. This explains the much faster hydrolysis rate of the amorphous PDLLA compared to semicrystalline PLLA. Vert et al. demonstrated, in a series of papers, that in the case of massive specimens of PLA or poly(lactide-co-glycolide) (PLG) the hydrolysis is faster in the center than at the surface [35, 36]. They suggested that the hydrolysis products are formed both at the surface and in the inner part, but those localized near the surface dissolve in the aging medium, while the concentration of carboxylic end groups increases in the center and catalyze ester hydrolysis, resulting in a surface-center segregation and multimodal molecular weight distributions for both amorphous and semicrystalline polymers. When compression-molded plates, millimetric beads, microspheres, and cast films were fabricated from the same batch and aged in phosphate buffer pH 7.4 and 37 °C, the plates and beads degraded faster, with bulk disintegration, compared to surface hydrolysis of the films and microspheres [51]. The heterogeneous degradation has also been observed during hydrolysis of lactide copolymers with 1,5-dioxepan-2-one [42] and glycine [52].

The composition of the polymer chains, i.e., the content of L-LA, D-LA, and/or copolymer units greatly influences the degradation rate of PLA polymers [37, 40, 53]. The copolymerization affects both the crystallinity and hydrophilicity of the polymer which have a large influence on the hydrolysis rate. The half-life of various PLLA, PDLLA, and PLG polymers allowed to degrade in pH 7.4 phosphate buffer at 37 °C varied from 3 to 110 weeks depending on the amount of L-LA, D-LA, and GA units in the polymer [53]. The semicrystalline PLLA containing 100% L-LA units had the longest degradation time with a half life of 110 weeks, the incorporation of 50% D-LA units greatly increased the degradation rate and the half life decreased to only 10 weeks [35]. The copolymerization with 25% GA further decreased the half life to only 3 weeks [40]. The half life corresponds to the time at which half of the initial material is lost.

The hydrolysis of homo- and copolymers of PLA and PGA proceeded in three stages [37]. During the first stage the molecular weights of the polymers decreased rapidly, probably through random-chain scission, but no significant weight loss occurred. In stage two the molecular weight decrease was followed by increasing mass loss and monomer formation. At the time of total weight loss about 50% of the copolymers had been hydrolyzed to monomer. During the third stage the hydrolysis of soluble oligomers continued until the polymer was totally hydrolyzed to the monomers, lactic and glycolic acids. The hydrolysis rate of pure PLLA was much lower and did not reach stage three with total mass loss until around 80 days of aging at 60 °C. The degradation rate of L-LA and DL-LA copolymers with DXO could be varied over a wide range by changing the copolymer composition [42]. Crystallinity was observed with up to 50% incorporation of DXO [54]. In the case of tri-block copolymers of poly(L-lactide-*b*-1,5-dioxepan-2-one-*b*-L-lactide) the rate of degradation in buffered solution at pH 7.4 and 37 °C was only influenced by the original molecular weight and not by the copolymer composition [55]. Due to the amorphous nature of PDXO, the DXO blocks were degraded faster than the L-LA blocks. Hydrolytic degradation of poly(caprolactone-*b*-L-lactide) block copolyesters at pH 7.4 and 37 °C was controlled by the initial crystallinity and overall composition [56]. The rate of degradation increased with increasing PLLA content.

Several studies have shown that the presence of certain enzymes, e.g., pronase, proteinase K, and bromelain increases the degradation rate of polylactide, while other enzymes, e.g., esterase and lactate dehydrogenase have no effect on the degradation rate [57, 58, 59]. When the enzymatic degradability of several polylactides with the L repeat unit content varying from 0 to 99% was studied, it was shown that proteinase K preferentially degraded PLLA compared to PDLLA. [58,59]. However, the high crystallinity of pure PLLA decreases the degradation rate [60]. The fastest weight loss and surface degradation were found for the polymer containing 92% L-units. In accordance, Li et al. showed that PLA50-*mes* degraded much faster than PLA50-*rac*. This means that proteinase K preferentially degrades L,L-, D,L-, and L,D-bonds as opposed to D,D ones. [61]. In another study,

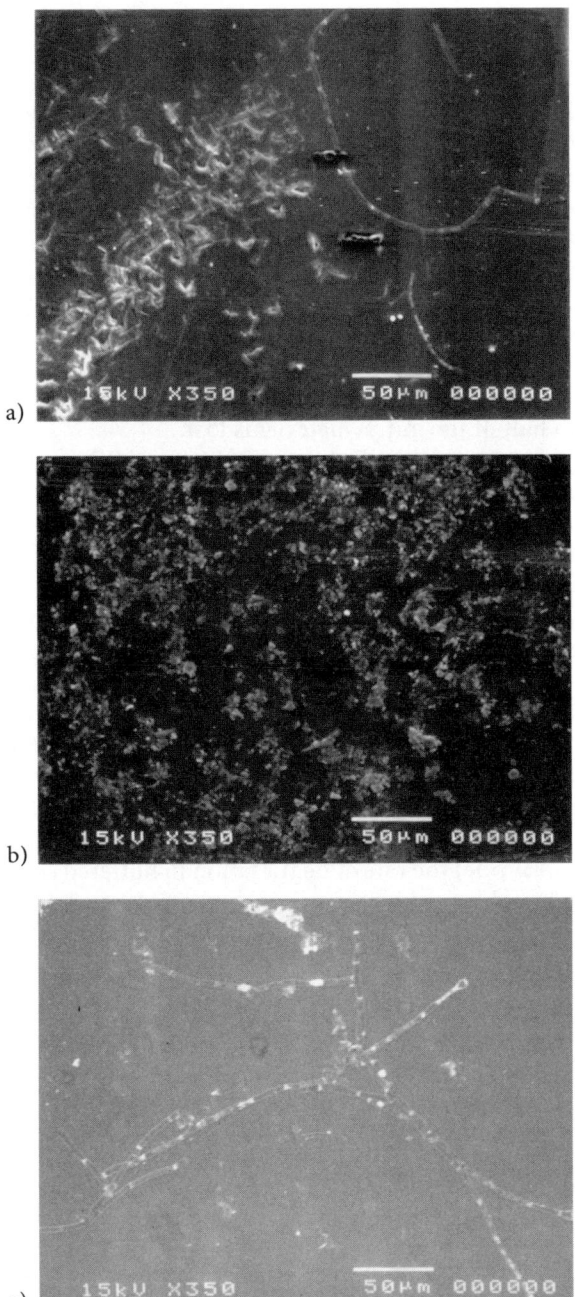

Fig. 1. SEM micrographs of biotically aged polylactide films: *a*) surface erosion, *b*) bacteria, and *c*) fungi on the surface of the PLLA films

a PLA stereopolymer containing 50% L-units showed the most rapid enzymatic degradation, when low molecular weight (Mn 2000) PLA stereopolymers were exposed to several esterase enzymes [62].

There has been only a few studies in which PLA or PLA oligomers were subjected to selected microorganisms [63–66]. In general, these studies state that the molecular weight decreased initially by abiotic hydrolysis, but after the initial abiotic decrease, the molecular weight for the samples aged in biotic medium decreased faster than the molecular weight of the samples aged in abiotic medium. Racemic PDLLA exposed to *Fusarium moniliforme* and *Pseudomonas putida* at 30 °C degraded initially through chemical hydrolysis followed by bioassimilation of the by-products. [63]. A mixed culture of *F. moniliforme* and *P. putida* resulted in faster assimilation of oligomers compared to pure cultures of *F. moniliforme* or *P. putida* [64].

The presence of compost microorganisms in the mineral medium accelerated the degradation of polylactide compared to the degradation in the corresponding sterile mineral medium [67]. After 5 weeks in biotic environment the films had fragmented to fine powder, while the films in a corresponding abiotic medium still looked intact. A rapid molecular weight decrease and increasing polydispersity was observed in the biotic environment. In the abiotic environment only a slight molecular weight decrease was seen and the polydispersity started decreasing towards 2. This indicates different degradation mechanisms, i.e., preferred degradation near the chain ends in the biotic environment and a random hydrolysis of the ester bonds in the abiotic environment. SEM micrographs showed surface erosion on the films aged in biotic medium (Fig. 1, *a*), while the surface of the sterile films remained smooth. A large number of bacteria and mycelium of fungi were also seen on the surface of the biotically aged films (Fig. 1, *b* and *c*). The presence of low molecular weight lactic acid derivatives in the films enhanced the degradation of polylactide in biotic medium [68].

Pranamuda et al. have investigated the distribution of PLA degrading microorganisms in soil [65]. They isolated a microbial strain with a high level of PLA degrading activity and demonstrated that 60% of PLA film was degraded by this *Amycolatopsis* sp. after 14 days in liquid culture. The film surfaces remained smooth in the sterile controls, but in the inoculated films the surfaces became rough and hemispherical holes were formed. The molecular weight of pure PDLLA films, buried in compost, was reduced to half of the original value after only 15 days [66]. The rapid molecular weight decrease was explained by the combined effect of thermal degradation, hydrolysis, and biodegradation taking place during composting.

2.2
Degradation of Polycaprolactone

Polycaprolactone is relatively stable against abiotic hydrolysis, but it was shown early that it is degraded by microorganisms [69]. The molecular weight of PCL remained unchanged during the degradation and it was stated that the biotic

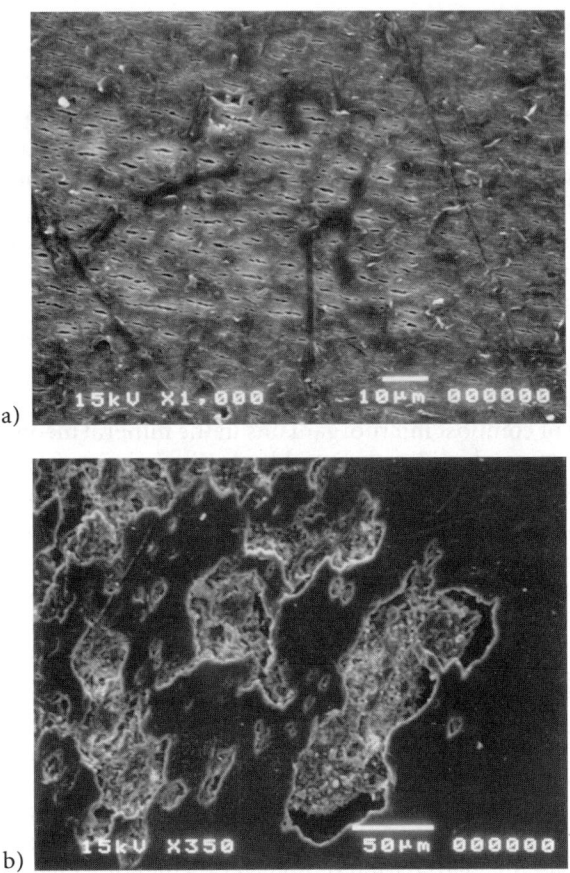

Fig. 2. SEM micrographs of biotically aged polycaprolactone films: *a*) parallel cracks and *b*) surface erosion on the PCL films

degradation occurred at the surface. Later, polycaprolactone has been shown to biodegrade in many different environments, e.g., in pure fungal cultures [70–73], in compost [73–76], in active sludge [73, 76], by enzymes [77], and in soil [78]. PCL degrading anaerobic microorganisms are found in several natural environments such as river water, sewage sludge supernatant, farm soil, paddy soil, creek sediment, roadside sand, and pond sediment [79, 80]. Generally, it has been shown that the biodegradation of PCL proceeds by rapid weight loss through surface erosion with minor reduction of the molecular weight. In contrast, the abiotic hydrolysis of PCL proceeds by a reduction in molecular weight combined with minor weight loss.

Incubation of PCL with *Aspergillus* sp., *Penicillium funiculosum*, *Chaetomium globosum*, and a *Fusarium* sp. showed that the degradation rate was controlled by molecular weight and degree of crystallinity [81]. The faster degradation of the amorphous regions by *Aspergillus flavus* and *Penicillium funiculosum* revealed a spherulitic pattern through the action of extra-cellular enzymes [82]. In another study surface erosion of PCL films in compost, in anaerobic sludge, and by *Aspergillus fumigatus* was compared with chemical hydrolysis at 23 °C and 50 °C [73]. Differences in microflora together with the initial morphology of the sample resulted in different mechanisms of erosion. The degradation in compost resulted in parallel grooves or cracks, while incubation with *A. fumigatus* produced a spherulitic erosion pattern. Accordingly, parallel grooves or cracks were formed in the films aged in mineral medium containing a mixed culture of compost organisms [83]. Fig. 2. shows some of the surface erosion patterns formed during aging with compost microorganisms. MALDI-TOF analysis showed that a preferential degradation in the amorphous parts produced low-molar mass fractions with lengths corresponding to one, two, or several times the thickness of the crystal lamellae [73]. Temperature played an important role in the degradation in compost and anaerobic sludge [84, 85]. Probably, the observed changes in degradation rate in biological environments at different temperatures are a result of changes in microflora [86].

It has been demonstrated that the erosion of PCL proceeds in the vicinity of chain-ends [74]. Accordingly, the presence of phthalic end groups as end-cappers reduced the biodegradability of PCL exposed to pure and mixed cultures of microorganisms [87]. Recycling or addition of processing additives slightly decreased the degradation rate compared to the degradation of pure PCL [85]. Cross-linking decreased the enzymatic degradation rate [88]. The unidirectional orientation of semicrystalline polycaprolactone films strongly affected the biodegradation rate [89]. After 700 h of incubation the weight loss was between 99 and 25% depending on the orientation.

The presence of starch significantly increased the biodegradation rate of PCL in activated sludge, soil burial, and controlled composting [90]. The amount of corn starch blended with PCL did not affect the degradation caused by *Rhizopus arrhizus* lipase. However, the degradation rate increased with increased starch content in the case of *Bacillus subtilis* α-amylase [91]. The biodegradation of PCL and poly(ethylene terephthalate) (PET) blends was tested in full-scale composting, soil burial, and under exposure to different fungi and esterases [17]. The biodegradation detected in the samples was well below what would be expected from the behavior of the homopolymers under similar environmental conditions. A thin layer of poly(vinyl alcohol) (PVAl) on the surface of PCL also prevented biodegradation [18]. The in vitro degradation of 1,3-dioxan-2-one (TMC) and caprolactone copolymers proceeded faster than the degradation of PTMC homopolymer, but more slowly than the hydrolysis of PCL homopolymer [92, 93]. Amorphous parts were degraded first leading to increased crystallinity. The M_w for the copolymer decreased from 120,000 to 40,000 during 450 days, while the molecular weight decrease for pure PTMC did not start until 600 days in vitro. The copolymerization with PCL, thus, considerably increased the hydrolysis rate.

2.3
Degradation of PHB and PHBV

The in vitro degradation of PHB and PHBV under physiological conditions is very slow [94]. Under accelerated conditions (high temperature and/or acidic/basic pH) the degradation proceeds through a molecular weight decrease and when the molecular weight is sufficiently low a weight loss is observed, almost all the mechanical strength is lost and the remaining polymer breaks down into small fragments [94,95]. The rate of chemical hydrolysis decreases with increasing crystallinity [96, 97]. There is no agreed explanation as to how the copolymer composition affects the hydrolysis rate [95,96]. It has been suggested that it is the crystallinity, rather than the composition that affects the hydrolysis rate [96]. On the other hand, PHBVs of the same crystallinity, but different compositions (45–71 mol % HV), showed decreasing hydrolysis rates with increasing HV content [95].

The bacterially produced poly(hydroxyalkanoates) are quite resistant to moisture, but they are rapidly biodegraded by a wide range of micro-organisms [98]. The rate of enzymatic degradation of PHB and PHBV by PHA depolymerase was from two to three orders of magnitude faster than the rate of simple hydrolytic degradation [95]. The enzymatic hydrolysis of PHB and PHV copolymers is a heterogeneous erosion process proceeding from the surface, where polymer chains are degraded initially by *endo*-scissions (randomly throughout the chain) and then by *exo*-scissions (from the chain ends) [99, 100, 101]. This results in subsequent surface erosion and weight loss. The average molecular weight and molecular weight distribution do not change during the enzymatic degradation because of the selective degradation only at the surface, together with the removal and dissolution of low-molecular weight degradation products from the polymer matrix into the surrounding environment. The preferential enzymatic attack of the amorphous phase of PHB has been reported [102]. Using ^{1}H-NMR imaging combined with total weight loss measurements, it was found that in the initial stages of degradation only amorphous material was consumed [103]. Later, however. both amorphous and crystalline regions were degraded without preference. The polymer structure affected the enzymatic degradation, e.g., a large number of HV units in the PHBV copolymer reduced the extent of enzymatic degradation [98, 104]. Long side-chain at the β-carbon further reduced the possibility of enzymatic attack [23].

PHB and PHBV degrade in several natural environments, such as soil [105], compost [106, 107], sea water [108], and anaerobic activated sludge [109]. Degradation is mainly characterized by a large weight loss, but also to some extent by molecular weight decrease [106, 108, 109]. During the composting of organic material, moisture is present and temperature may reach 60–80 °C. This means that both hydrolytic and thermolytic processes are accelerated and they may play an important role in the overall degradation of the polymer. The effect of abiotic factors, such as water and air, on the degradation of PHBV in garden waste compost has been investigated using simulated and natural environments

[110, 111]. These included exposure to sterile water or air at 60 °C, pure fungal cultures and garden waste compost. This study showed that during a period of 50 days, water and air had very little or no effect on the degradation of PHVB in garden waste compost. The degradation was due to microbial action only.

3
Degradation Products of Aliphatic Polyesters

Degradation of polymers results in a mixture of degradation products varying in chemical composition and molecular weight. Some of the degradation products remain in the material, while others are able to migrate out to the surrounding environment, e.g., into soil, water, or air. Identification of these low molecular weight products is important to fully understand the degradation mechanism and interaction of degradable polymers with the environment or the human body. Degradation of some polymers, e.g., polyethylene leads to the formation of complex product mixtures containing up to hundreds of different products [112, 113, 114], while other polymers such as polyesters contain functional groups in their main chain where degradation is initially thought to take place. The resulting products are less complex and more predictable [37, 110]. New techniques need to be developed to identify the low molecular weight degradation products. These methods comprise methods for isolation of the products from the polymeric material and/or the degradation media (e.g., water, salt solution, soil, air), preconcentration of the compounds, methods for further separation, and finally identification. The nature of the matrix and the products to be extracted determine the choice of the extraction method. An ideal extraction method is quantitative, selective, rapid, and uses little or no solvent.

Liquid-liquid extraction (LLE) is the traditional method to extract organic compounds from water. The low molecular weight compounds are transferred from one liquid phase to another immiscible or partially immiscible liquid by shaking them in a separation funnel. LLE is still a common method, but has several drawbacks such as low selectivity, labor intensivity, and the use of large amount, of organic solvent. LLE has been used to extract hydrolysis products of degradable polyesters such as PLA and its copolymers from the buffer solution [115].

Solid phase extraction (SPE) is a more sophisticated method for separation of volatile and semivolatile substances from water or organic solvent [116, 117]. It also allows the separation of the products into several fractions, which is preferable with complex product patterns. The principle of SPE is analogous to that of LLE, but it gives highly selective extractions with a minimum of solvent. During the extraction a liquid sample is passed over a solid or "sorbent" that is packed in a medical-grade polypropylene cartridge. As a result of strong attractive forces between the isolates and the sorbent, the isolates are retained on the sorbent. Later, the sorbent is washed with a small volume of a solvent that has the ability to disrupt the bonds between the isolates and the sorbent. The isolates are concentrated in a relatively small volume of clean solvent and are ready to be injected into a GC. The selectivity of the extraction is a function of the chemical structure of the isolate, the properties of the solid phase, and the composition of the sample matrix. The specific properties of a given bonded silica sorbent are a result of the functional group covalently bonded to the silica substrate. By using the SAX ion-exchange column we have demonstrated a complete conversion of PLG copolymers to lactic acid and glycolic acid [37]. The results also showed that the hydrolysis rate of glycolide units was faster than that of lactide units giving increasing content of lactide units in the remaining copolymers. Solid-phase extraction with ENV+ columns and subsequent GC-MS analysis showed different product patterns after biotic and abiotic hydrolysis of PLLA [67, 68] and PHBV [110, 111].

Headspace-GC-MS analysis is useful for the determination of volatile compounds in samples that are difficult to analyze by conventional chromatographic means, e.g., when the matrix is too complex or contains substances that seriously interfere with the analysis or even damage the column. Peak area for equilibrium headspace gas chromatography depends on, e.g., sample volume and the partition coefficient of the compound of interest between the gas phase and matrix. The need to include the partition coefficient and thus the sample matrix into the calibration procedure causes serious problems with certain sample types, for which no calibration sample can be prepared. These problems can, however, be handled with multiple headspace extraction (MHE) [118]. Headspace-GC-MS has been used for studying the volatile organic compounds in polymers [119]. The degradation products of starch/polyethylene blends [120] and PHB [121] have also been identified.

The solid phase microextraction technique was developed by Pawliszyn and coworkers [122]. It is an inexpensive, rapid, and solvent-free technique for the isolation of organic compounds from air, water, and soil. We have also demonstrated the usefulness of SPME in extracting degradation products from polymers [123–125]. SPME is based on a typically one-centimeter long, thin, fused silica fiber coated with a polymeric stationary phase mounted in a modified syringe. For sampling. the fiber is immersed directly into aqueous samples or in the headspace over the liquid or solid sample matrix. The organic compounds are absorbed on the fiber and later they are thermally desorbed in the injection port of GC and transferred to the capillary column by a carrier gas. SPME is an equilibrium technique, therefore the organic compounds are not completely extracted from the matrix. Nevertheless, the method is useful for quantitative work and excellent precision and linearity have been demonstrated.

3.1
Degradation Products of PLA and its Copolymers

Gas chromatography-mass spectrometry (GC-MS), liquid chromatography (LC), enzymatic assays, and capillary zone electrophoresis (CZE) have been used to monitor abiotic and biotic degradation products of polylactide and its copolymers [35, 37, 115, 126–128]. In abiotic aqueous environments the degra-

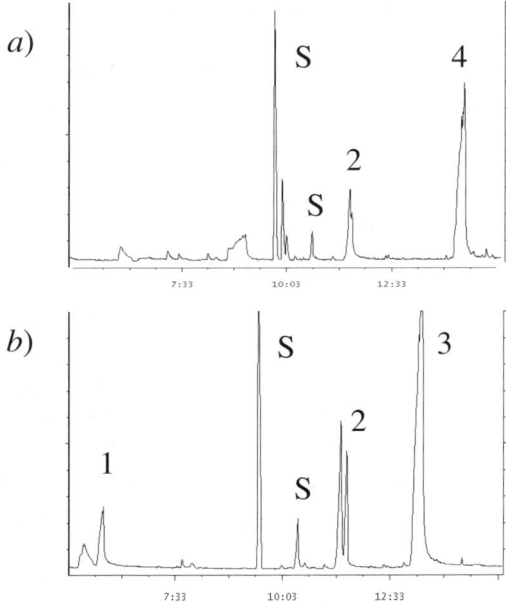

Fig. 3. GC-MS analysis showed different product patterns after aging PLLA in biotic and abiotic mineral medium. After 4 weeks in *a*) biotic mineral medium and *b*) sterile mineral medium. Peak identity: Peak *1*=lactic acid; peak *2*=lactide; peak *3*=lactoyllactic acid and peak *4*=ethyl ester of lactoyllactic acid

dation proceeds through hydrolysis of the ester bonds. This leads from the high molecular weight polymer to the intermediate degradation products (insoluble and soluble oligomers) and, finally, lactic acid is formed as the ultimate degradation product of abiotic hydrolysis [37].

Figure 3 shows the GC-MS chromatograms of the different product patterns obtained after PLLA was aged in abiotic and biotic mineral media [67, 68]. The biotic medium contained a mixed culture of compost microorganisms. The new biotic degradation products acetic acid, propanoic acid, and the ethyl ester of lactoyllactic acid were extracted after aging in biotic medium. Acetic acid and propanoic acid were formed as intermediate degradation products during the initial stages, but they were no longer detected after prolonged aging. The concentration of the ethyl ester of lactoyllactic acid, however, increased with aging time. Acetic acid, propanoic acid, and the ethyl ester of lactoyllactic acid were not detected after aging in sterile mineral medium, instead lactide, lactic acid, and lactoyllactic acid were identified. In the biotic medium lactic acid and lactoyllactic acid formed by abiotic hydrolysis were rapidly assimilated by the microorganisms, but a small amount of cyclic lactide remained detectable in the films.

Vert et al. exposed compression-molded racemic PDLLA to a mixed culture of *Fusarium moniliforme* and *Pseudomonas putida* at 30 °C [63]. The polymer initially degraded through chemical hydrolysis followed by bioassimilation of the by-products. HPLC analysis showed that DL- and L-lactic acids and L,L-dimer were rapidly assimilated by *F. moniliforme* and *P. putida*, whereas the assimilation of D,D-dimer proceeded slowly [64]. A mixed culture of *F. moniliforme* and *P. putida* resulted in faster assimilation of oligomers compared to pure cultures of *F. moniliforme* or *P. putida*. Pyrolysis of biotically and abiotically aged PLLA showed that the ratio of *meso*-lactide to L-lactide was lower in samples aged in biotic media compared to samples aged in abiotic media [129]. This confirms the preference of the microorganisms for the L-form of lactic acid and lactoyl-lactic acid.

Lactic acid and glycolic acid were identified by GC-MS after hydrolysis of PLG copolymers at pH 7.4 and 37 °C or 60 °C [37]. When the total weight loss occurred about 50% of the polymer had been converted to lactic and glycolic acid, afterwards the hydrolysis of soluble oligomers continued until a complete conversion of PLG copolymers to lactic acid and glycolic acid was demonstrated. Due to the higher hydrolysis rate of the glycolide units the lactide content of the copolymers increased during the hydrolysis. Lactic acid and 2-hydroxyethoxy-propanoic acid were the in vitro hydrolysis products of L-LA or D,L-LA and DXO copolymers at pH 7.4 and 37 °C [115]. The main degradation products formed during hydrolysis of poly(L-LA-*b*-DXO-*b*-L-LA) were lactic acid and 3-(2-hydroxyethoxy)-propanoic acid [55]. The release of degradation products started after 5 weeks which coincided with the onset of weight loss. In another study monomers and several other peaks assigned as oligomers up to the hexamer were identified by HPLC after alkaline hydrolysis of PLA and PLG copolymers [130].

In several papers enzymatic assays have been used to follow the release of L-lactic acid during hydrolysis of PLA and its copolymers [35, 36, 40, 51, 131, 132]. The major limitation with this method is that it only detects the L-form of lactic acid and does not give any information about the amount of D-lactic acid, oligomers, or other low molecular weight products. After hydrolysis of massive PDLLA specimens at pH 7.4 and 37 °C traces of lactic acid were first detected after 5 weeks, but most of the lactic acid was formed between 8th and 11th weeks, in agreement with weight loss [35]. In the case of semicrystalline PLLA L-lactic acid was not detected until week 31 [36]. During degradation in phosphate buffer at pH 7.4 and 37 °C for 28 days, the release of lactic acid increased as the percentage of the low molecular weight component in the blend was increased [132].

Vert et al. have, in a series of papers, demonstrated the use of capillary zone electrophoresis to analyze water-soluble oligomers formed during hydrolysis of PLA and its copolymers [133, 134, 135, 136]. By using CZE water-soluble oligomers of D,L-lactic acid with a degree of polymerization lower than 8 were detected in the buffer solution [133]. In another study CZE was used to monitor the in vitro aging of aqueous solutions of lactic acid oligomers [134]. The results showed that degradation of oligo(D,L-lactic acid) did not yield as much monomer as would be expected from purely random degradation. Ester scission of large oligomers formed predominantly the dimer. The authors concluded that the ester bond of lactoyllactic acid is more stable than the ester bonds inside the oligomer chains, thus giving the monomer at a slower rate. CZE was also used to complement the earlier studies on in vitro degradation of large-size lactic acid plates, showing that lactic acid monomer was the only component released into the aqueous media during the aging [134]. This indicates that either the higher oligomers are degraded inside the matrix or that water-soluble oligomers were released in the aging medium, but they were further degraded too rapidly to be detected. CZE analysis showed that after hydrolytic degradation of gluconic/glycolic/lactic acid, the copolymer returned to its ultimate building blocks, namely gluconic, glycolic, and lactic acids [136].

3.2
Degradation Products of PCL and its Copolymers

GC-MS chromatograms in Fig. 4 show that the low molecular weight PCL products were rapidly utilized by compost microorganisms. The unaged PCL films contained some low molecular weight products, i.e., 6-hydroxyhexanoic acid, caprolactone, and its cyclic dimer and trimer (Fig. 4, *a*). These products were also identified after aging in sterile mineral medium. However, after 2 weeks in

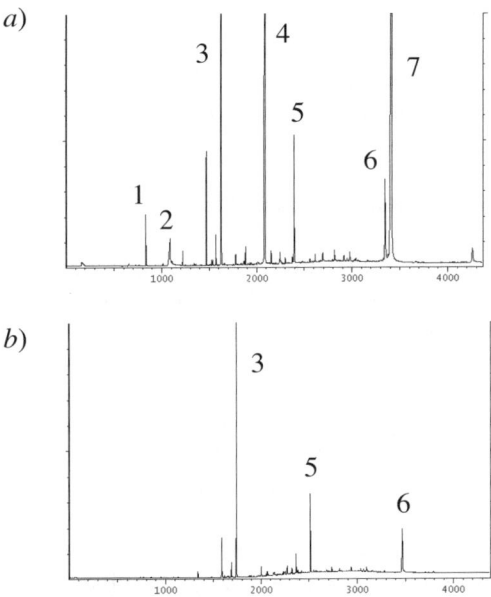

Fig. 4. GC-MS chromatograms of low molecular weight products extracted from PCL *a*) before aging and *b*) after 2 weeks in biotic mineral medium. The identity of the numbered peaks is: Peak *1*=caprolactone; peak *2*=6-hydroxyhexanoic acid; peak *4*=cyclic dimer; peak *7*=cyclic trimer; peaks *3*, *5* and *6*=phthalates

biotic medium, the low molecular weight products had been assimilated by the microorganisms (Fig. 4, *b*). MALDI-TOF analysis showed the formation of oligomers after biotic hydrolysis of PCL [73]. Succinic acid, butyric acid, valeric acid, and hexanoic acid were identified after biotic hydrolysis of PCL at 50 °C by *Aspergillus* sp. [137]. Enzymatic degradation of copolymers of 3-hydroxybutyric acid (3HB) and 6-hydroxyhexanoic acid (6HH) by PHB depolymerase from *A. faecalis* resulted in the formation of 3HB monomer, 3HB-3HB dimer, 6HH-3HB dimer, and 3HB-6HH-3HB trimer [23]. 6HH monomer and 6HH-6HH dimer were formed when the same copolymer was degraded by a lipase from *R. delemar*. The results suggested that the PHB depolymerase was incapable of hydrolyzing the ester bonds of 6HH-3HB and 6HH-6HH sequences in dimers and oligomers, while the lipase could only hydrolyze the ester bonds of the 6HH-6HH sequences. In another study the biodegradation of PHB/PCL blends resulted in the formation of PHB oligomers, but no degradation products of the PCL part could be identified [138]. The only degradation product identified after hydrolysis of poly(trimethylene carbonate-*co*-caprolactone) was 1,3-propanediol formed by the hydrolysis of the TMC regions of the polymer [93]. The absence of 6-hydroxyhexanoic acid or its oligomers was explained by the low hydrolysis rate of the CL regions. HPLC analysis of degradation products of a model drug polymer synthesized from 1,6-hexane diisocyanate (HDI), polycaprolactone diol (PCL), and the fluoroquinolone antibiotic, ciprofloxacin, showed the release of multiple degradation products including ciprofloxacin bonded to PCL fragments [139]. The polymer was incubated with solutions of cholesterol esterase (CE) or phosphate buffer (pH 7.0) at 37 °C. In addition to HPLC and GC-MS, CZE has been shown to be a convenient tool to monitor 6-hydroxyhexanoic acid and its oligomers in aqueous solutions [135].

3.3
Degradation Products of PHB and its Copolymers

The degradation products of PHB and its copolymers have been analyzed by GC [110, 140], HPLC [22, 23, 141–143], CZE [144], and NMR [98]. Solid-phase extraction and subsequent GC-MS analysis showed that the product patterns differ after biotic and abiotic hydrolysis of PHBV copolyesters [110]. During the degradation of PHBV by *Aspergillus fumigatus* (fungi) at 25 °C, 3-hydroxybutyric acid, 3-hydroxyvaleric acid, 3-hydroxybutyrate dimer, 3-hydroxybutyrate-3-hydroxyvalerate dimer, and 3-hydroxyvalerate dimer were detected after 10 days. The relative amount of 3-hydroxybutyric acid released was slightly higher than that of 3-hydroxyvaleric acid and in 21 days the polymer was completely degraded. When biotic hydrolysis was continued 3-hydroxybutyric acid and 3-hydroxyvaleric acid disappeared and acetic, butyric, and valeric acids were detected instead. At the same time, the mineral medium changed from transparent to yellow due to the excretion of these metabolites. In the sterile control no degradation products were detected during the 21-day experimental period. Similar degradation products were observed during the degradation of PHB by *Bacillus polymyxa* and *Bacillus megaterium* under oxygen-depleted conditions [145]. Water-soluble degradation products, mainly monomers and dimers, were formed during the enzymatic degradation of PHB and PHBV by PHA depolymerase [146, 147, 148]. The stereochemistry of the PHB polymer influenced the degradation product composition [22]. The enzymatic hydrolysis products of P[(*R*)-3HB] were the monomer and dimer, but chemosynthetic P(3HB) containing 4–20% (*S*)-3HB gave monomer, dimer, trimer, and tetramer as degradation products. This suggests that PHB depolymerase was incapable of hydrolyzing the ester bonds in (*S*)-3HB units. The rate of enzymatic hydrolysis of the (*R*)-3HB dimer was slower than the hydrolysis of higher oligomers. The distribution of water-soluble products after enzymatic degradation of poly(3-hydroxybutyrate-*co*-3-hydroxyhexanoate) suggested that the enzyme was incapable of hydrolyzing the ester bond of 3HB-3HH in the dimer and oligomers [149]. Accordingly, during enzymatic degradation of poly(3-hydroxybutyrate-*co*-6-hydroxyhexanoate) the PHB depolymerase from *A. faecalis* was incapable of hydrolyzing the ester bonds of 6HH-3HB and 6HH-6HH sequences [23]. The enzymatic degradation resulted in the formation of 3HB monomer, 3HB-3HB dimer, 6HH-3HB dimer, and 3HB-6HH-3HB trimer. No 6HH monomer or 6HH-6HH dimer was detected.

During aging of PHVB in sterile water at pH 7 and 60 °C, 2-butenoic acid, 2-pentenoic acid, 3-hydroxybutyric acid, 3-hydroxyvaleric acid, 3-hydroxybutyrate dimer, 3-hydroxybutyrate-3-hydroxyvalerate dimer, and 3-hydroxyvalerate dimer were formed [110]. The weight loss in sterile water was, however, slow, only 2% during 200 days at 60 °C. During thermal aging PHB and PHVB degrade through a six-membered transition state that produces 2-butenoic acid and 2-pentenoic acid and oligomers with butenoyl and pentenoyl end groups [150]. It has been shown that even at rather low temperatures such as 100 °C, 2-butenoic acid and 2-pentenoic acid are formed [121]. During processing the temperature varies between 150 and 170 °C [151]. The unsaturated products observed during hydrolysis in sterile water at 60 °C were thus formed as a result of hydrolytic cleavage of butenoyl and pentenoyl chain ends produced at the high temperatures used during processing.

The alkaline hydrolysis products of PHB and PHBV have been detected by LC [141, 143]. Monomers, oligomers, and derivatives, produced by dehydration at the OH-terminus were identified after the hydrolysis. According to these studies CZE showed that the accelerated hydrolysis of PHB led to the formation of hydroxyacid oligomers and a series of peaks formed by a side-reaction leading to the C=C bond at the non-carboxylic acid end [135].

4
The Influence of Processing and Processing Additives

There are rather few studies on how processing and processing additives affect the properties and degradation of aliphatic polyesters. Several types of processing additives are used in order to improve the properties of the product. Since these additives change the characteristics of the polymer they could change the susceptibility to (bio)degradation, e.g., by changing the surface properties.

4.1
The Influence of Processing on the Properties of Polylactide

PLA is sensitive to high temperatures and exhibits rapid loss of molecular weight during processing [152]. The film-blowing of PLA significantly reduced the molecular weight, e.g., M_w was reduced by 50%. During melt-pressing and melt-extrusion pure, PLA and PLA with SiO_2 exhibited the largest reductions in molecular weight, while $CaCO_3$ increased the thermal stability of PLA during processing. The reduction in M_w during melt-extrusion ranged from 35% to 50% depending on the extrusion temperature [152, 153]. Migliaresi et al. studied the effect of thermal treatment on crystallization and molecular weight of compression-molded PLLA [154]. They showed that the molecular weight was reduced between 48 to 72% after compression-molding at 210 °C. During melt-spinning of PLLA the viscosity-average molecular weight was reduced by 70% [155]. The molecular weight of PLA decreased between 50 to 88% during injection molding at different temperatures [156].

The influence of processing additives, i.e., erucamide (ER), silicon dioxide (SiO_2), and calcium carbonate ($CaCO_3$) on biotic and abiotic degradation of melt-pressed PDLLA films has been investigated [66]. ER is a primary fatty acid amide used as a slip agent. SiO_2 and $CaCO_3$ are inorganic compounds used as anti-blocking agents to roughen the surface of a polymer. The addition of $CaCO_3$ to PLA significantly reduced the thermal degradation during processing. However, it also retarded the biodegradation rate compared to PLA without additives and PLA with SiO_2. $CaCO_3$ neutralizes carboxylic end groups formed during hydrolysis or thermal degradation [157]. Without the catalytic effect of carboxylic acid groups the degradation rate is reduced. The low molecular weight citrate ester plasticizers increased the rate of enzymatic degradation compared to that of unplasticized PLA [158].

4.2
The Influence of Processing on the Properties of Polycaprolactone

PCL shows relatively high thermal stability compared to PLA and PHBV and can be processed without significant molecular weight reduction by, e.g., film-blowing, injection molding, and sheet extrusion. For example, during film-blowing of PCL the reduction in molecular weight was less than 10% compared to the molecular weight of the granules [152]. During reprocessing (recycling) the molecular weight of PCL decreased by about 20% [85]. The recycled films exhibited slightly lower tensile modulus and stress at yield and higher strain at yield, strain at break, and stress at break. During composting the molecular weight of virgin PCL decreased faster than the molecular weight of recycled PCL.

The addition of processing additives: erucamide, SiO_2, or erucamide+$CaCO_3$ in the film-blown PCL resulted in a slightly lower biodegradation rate compared to PCL without additives [85]. However, the addition of SiO_2 to melt-pressed PCL films increased the biodegradation rate [66]. The use of a slip additive, erucamide, did not affect the degradation rate of the melt-pressed films. The biotic degradation was in both cases studied in compost and in mineral medium with *A. fumigatus*. The biodegradability of PCL fibers with different draw ratios decreased with increasing draw ratio and crystallinity [159]. Studies of blown and uniaxially drawn aliphatic copolyesters exposed to *Penicillium pinophilum* and *Aspergillus niger*, also showed that the rate of biodegradation was reduced by increasing orientation [160, 161].

4.3
The Influence of Processing on the Properties of PHB and PHBV

PHB and PHBV have limited stability at the high temperatures involved in melt processing. It has been shown that the molecular weight is reduced through random scission of ester bonds at high temperatures [162]. PHBV is considered to be thermally unstable at temperatures above 170 °C and thermally stable below 160 °C. The molecular weight of PHBV showed great variations depending on the chosen processing conditions during melt extrusion [151]. The weight-average molecular weight decreased with increasing processing temperature or decreasing extrusion rate (higher residence time of melt inside the extruder). At a residence time of 12 minutes (5 rpm) and an extrusion temperature of 177 °C, the M_w decreased to 50% of the original value. Similar results were obtained after injection molding and melt pressing of PHBV, where the M_w decreased by 50% and 60%, respectively [163, 164]. An increase in HV content led to a smaller reduction in molecular weight, because milder processing conditions could be utilized. In addition to molecular weight, tensile strength and elongation at break decreased with increasing processing temperature and time [163]. SEM micrographs showed that the surface of the samples processed at low temperature was smooth, while the samples processed at high temperature and low extrusion rate exhibited small pits at the surface due to the thermal degradation [151]. The degree of crystallinity remained practically unchanged during processing, but X-ray and DSC results showed that processing affected the atomic and morphological structure of PHBV [165].

The degradation rate of melt-extruded PHBV in pure fungal cultures (*Phanerochaete chrysosporium*, *Penicillium simplicissimum*, and *Aspergillus fumigatus*) and by chemical hydrolysis at 60 °C and pH 10.5 increased for the samples processed at high temperature [166]. When samples of PHBV prepared by different processing methods were subjected to hydrolytic degradation at different pH and temperature conditions, the cold-pressed and solvent-cast films had the lowest and the melt-pressed and injection molded the highest hydrolytic stability [96, 97]. The differences were attributed to differences in degree of compaction and crystallinity. The rate of hydrolysis increased with decreasing molecular weight. The influence of thermal history on enzymatic degradation of PHBV has also been investigated [167]. Annealed melt-cast samples with high degree of crystallinity, crystalline perfection, and spherulite size showed lower rates of degradation. The influence of the processing method on the degradation rate of PHBV exhibited the following order: solution-cast film>extruded film>surface of injection-molded film>core of injection-molded film.

5
Conclusions

There are many similarities, but also many differences, in the degradation behavior of the aliphatic polyesters. Large variations in the properties and degradation sensitivity are obtained by copolymerization and blending. The sensitivity to moisture increases in the order polycaprolactone, poly(3-hydroxybutyrate) and polylactide. Poly(3-hydroxybutyrate) shows the fastest biodegradation rate, while polycaprolactone has the highest thermal stability, e.g., during processing. Different degradation product patterns are obtained after aging in different biotic and abiotic environments. The stereochemistry of the PHB, PHBV, and PLA polymers has a large effect on the biodegradation rate. It also influences the formation and assimilation of the low molecular weight degradation products.

Acknowledgement. Professor Ann-Christine Albertsson is thanked for valuable discussions.

References

1. Kalb B, Pennings AJ (1980) Polymer 21:607
2. Gilding DK, Reed AM (1979) Polymer 20:1459
3. Grijpma DW, Nijenhuis AJ, Pennings AJ (1990) Polymer 31:2201
4. Fukuzaki H, Yoshida M, Asano M, Aiba Y, Kaetsu I (1990) Eur Polym J 24:1029
5. Karjalainen T, Hiljanen-Vainio M, Seppälä J (1996) J Appl Polym Sci 59:1289
6. Lostocco, MR, Murphy CA, Cameron JA, Huang SJ (1998) Polym Degrad Stab 59:303
7. Löfgren A, Albertsson A-C (1994) J Macromol Sci Pure Appl A 52:1327
8. Stridsberg K, Albertsson A-C (2000) J Polym Sci Polym Chem 38:1774
9. Buchholz B (1993) J Mater Sci Mater Med 4:381
10. Mauduit J, Perouse E, Vert M (1996) J Biomed Mater Res 30: 201
11. Hiljanen-Vainio M, Varpomaa P, Seppälä J, Törmälä P (1996) Macromol Chem Phys 197:1503
12. Tsuji H, Mizuno A, Ikada Y (1998) J Appl Polym Sci 70:2259
13. Focarete ML, Ceccorulli G, Scandola M, Kowalczuk M (1998) Macromolecules 31:8485
14. Lunt J (1998) Polym Degrad Stab 59:145
15. Sinclair RG (1996) J Macromol Sci Pure Appl Chem A 33:585
16. Coleman MM, Zarian J (1979) J Polym Sci 17:837
17. Chiellini E, Corti A, Giovanni A, Narducci P, Paparella AM, Solaro R (1996) J Environ Polym Degrad 4:37
18. DeKesel C, Vander Wauven C, David C (1997) Polym Degrad Stab 55:107
19. Tilstra L, Johnsonbaugh D (1993) J Environ Polym Degrad 1:247
20. Pranamuda H, Tokiwa Y, Tanaka H (1996) J Environ Polym Degrad 4:1

21. Kister G, Cassanas G, Bergounhon M, Hoarau D, Vert M (2000) Polymer 41:925
22. Abe H, Doi Y, Kumagai Y (1994) Macromolecules 27:6012
23. Abe H, Doi Y, Aoki H, Akehata T, Hori Y, Yamaguchi A (1995) Macromolecules 28:7630
24. Albertsson A-C, Gruvegård M (1995) Polymer 36:1009
25. Shalaby S, Kafrawy A (1989) J Polym Sci Polym Chem 27:4423
26. Löfgren A, Albertsson A-C, Dubois Ph, Jerome R, Teyssie Ph (1994) Macromolecules 27:5556
27. Lemoigne M (1925) Ann Inst Past 39:144
28. Doi Y, Tamaki A, Kunioka M, Soga K (1988) Appl Microbiol Biotechnol 28:330
29. Kawaguchi Y, Doi Y (1992) Macromolecules 25:2324
30. Amor SR, Rayment T, Sanders JKM (1991) Macromolecules 24:4583
31. Kawaguchi Y, Doi Y (1990) FEMS Microbiol Lett 70:151
32. Holmes PA (1985) Phys Technol 16:32
33. Ramsay JA, Ramsay BA (1990) Appl Phys 7:1
34. Hobbs JK, Barham PJ (1997) Polymer 38:3879
35. Li SM, Garreau H, Vert M (1990) J Mater Sci Mater Med 1:123
36. Li SM, Garreau H, Vert M (1990) J Mater Sci Mater Med 1: 198
37. Hakkarainen M, Albertsson A-C, Karlsson S (1996) Polym Degrad Stab 52:283
38. Reed AM, Gilding DK (1981) Polymer 22:494
39. Grijpma, DW, Pennings AJ (1994) Macromol Chem Phys 195:1633
40. Li SM, Garreau H, Vert M (1990) J Mater Sci Mater Med 1:131
41. Cam D, Hyon S-H, Ikada Y (1995) Biomaterials 16:833
42. Albertsson A-C, Löfgren A (1994) J Appl Polym Sci 52:1327
43. Hyon S-H, Jamshidi K, Ikada Y (1998) Polym Int 46:196
44. Zhang X, Wyss UP, Pichora D, Goosen MFA (1994) Bioactive and Compatible Polymers 9:80
45. Södergård A, Selin J-F, Näsman JH (1996) Polym Degrad Stab 51:351
46. Pierre St, Chiellini E (1987) J Bioact Compat Polym 2:4
47. Pennings JP, Dijkstra H, Pennings AJ (1993) Polymer 34:942
48. Vainionpää S, Rokkanen P, Törmälä P (1989) Prog Polym Sci 14:679
49. Fischer EW, Sterzel HJ, Wegner G (1973) Polymer 251:980
50. Chu CC (1981) J Appl Polym Res 15:1727
51. Grizzi U, Garreau H, Li S, Vert M (1995) Biomaterials 16:305
52. Helder J, Dijkstra PJ, Feijen J (1990) J Biomed Mater Res 24:1005
53. Li S (1999) J Biomed Mater Res (Appl Biomater) 48:342
54. Gruvegård, Lindberg T, Albertsson A-C (1998) J Macromol Sci Pure Appl Chem A 35:885
55. Stridsberg K, Albertsson A-C (2000) Polymer 41:7321
56. Ye WP, Du FS, Zin WH, Yang Y, Xu Y (1997) React Funct Polym 32:161
57. Williams DF (1981) Engineering in Medicine 10:5
58. Reeve S, McCarthy SP, Downey MJ, Gross RA (1994) Macromolecules 27:825
59. MacDonald RT, McCarthy SP, Gross RA (1996) Macromolecules 29:7356
60. Cai H, Dave V, Gross RA, McCarthy SP (1996) J Polym Sci Part B Polym Phys 34:2701
61. Li S, Tenon M, Garreau H, Braud C, Vert M (2000) Polym Degrad Stab 67:85
62. Fukuzaki H, Yoshida M, Asano M, Kumakura M (1989) Eur Polym J 25:1019
63. Torres A, Li SM, Roussos S, Vert M (1996) J Appl Polym Sci 62:2295
64. Torres A, Li SM, Roussos S, Vert M (1996) J Environ Poly Degrad 4:213
65. Pranamuda H, Tokiwa Y, Tanaka H (1997) Appl Environ Microbiol 63:1637
66. Renstad R, Karlsson S, Sandgren Å, Albertsson A-C (1998) Environ Polym Degrad 6:209
67. Hakkarainen M, Karlsson S, Albertsson A-C (2000) Polymer 41:2331
68. Hakkarainen M, Karlsson S, Albertsson A-C (2000) J Appl Polym Sci 76:228
69. Fields RD (1973) PhD Thesis, USA
70. Fields RD, Rodriguez F, Finn RK (1974) J Appl Polym Sci 18:3571

71. Murphy CA, Oda JAY, Asari H, Urakami T, Tonomura K (1995) J Ferment Bioeng 80:265
72. Cameron JA, Huang SJ, Vinopal RT (1996) Appl Environ Microbiol 62:456
73. Eldsäter C, Erlandsson B, Renstad R, Albertsson A-C, Karlsson S (2000) Polymer 41:1297
74. Lefebvre F, David C, Vander Wauven C (1994) Polym Degrad Stab 45:347
75. Ohtaki A, Akakura N, Nakasaki K (1998) Polym Degrad Stab 62:279
76. Bastioli C, Cerutti A, Guanella I, Romano GC, Tosin M (1995) J Environ Polym Degrad 3:81
77. Gan Z, Liang Q, Zhang J, Jing X (1997) Polym Degrad Stab 56:209
78. Toncheva V, Van Den Bulcke A, Schacht E, Mergaert J, Swing J (1996) J Environ Polym Degrad 4:71
79. Nishida H, Tokiwa Y (1994) Chem Letters 1293
80. Doi Y, Kasuya K-I, Abe H, Koyama N, Ishiwatari S-I, Takagi K, Yoshida Y (1996) Polym Degrad Stab 51:281
81. Benedict CV, Caeron JA, Huang SJ (1983) J Appl Poly Sci 28:335
82. Cook WJ, Cameron JA, Bell JP, Huang SJ (1981) J Polym Sci Polym Lett Ed 19:159
83. Hakkarainen M, Albertsson A-C (2001) manuscript in preparation
84. Pettigrew CA, Reece GA, Smith MC, King LW (1995) J Macromol Sci Pure Appl Chem A 32:811
85. Albertsson A-C, Renstad R, Erlandsson B, Eldsäter C, Karlsson S (1998) J Appl Poly Sci 70:61
86. Mergaert J, Webb A, Andersson C, Wouters A, Swings J (1993) J Appl Environ Microbiol 59:3233
87. Benedict CV, Cook WJ, Jarrett P, Cameron JA, Huang SJ, Bell JP (1983) J Appl Polym Sci 28:327
88. Darwis D, Mitomo H, Enjoji T, Yoshii F, Makuuchi K (1998) Polym Degrad Stab 62:259
89. Lefebvre C, David C, Villers D (1999) Makromol Chem Phys 200:1374
90. Bastioli C, Innocenti FD, Guanella I, Romano G (1995) JMS Pure Appl Chem A32:839
91. Tokiwa Y, Iwamoto A, Koyama M (1990) Polym Mater Sci Eng 63:742
92. Albertsson A-C, Eklund M (1994) J Polym Sci Part A Polym Chem 32:265
93. Albertsson A-C, Eklund M (1995) J Appl Poly Sci 57:87
94. Kanesawa Y, Doi Y (1990) Makromol Chem Rapid Commun 11:679
95. Doi Y, Kanesawa Y, Kunioka M, Saito T (1990) Macromolecules 23:26
96. Holland SJ, Jolly AM, Yasin M, Tighe BJ (1987) Biomaterials 8:289
97. Yasin M, Holland SJ, Tighe BJ (1990) Biomaterials 11:451
98. Kanesawa Y, Tanashi N, Doi Y, Saito T (1994) Polym Degrad Stab 45:179
99. Hocking PJ, Marchessault RH, Timmins MR, Lenz RW, Fuller RC (1996) Macromolecules 29:2427
100. Scherer TM, Fuller RC, Lenz RW (1994) J Environ Polym Degrad 2:263
101. Iwata T, Doi Y, Tanaka T, Akehata T, Shiromo M, Teramachi S (1997) Macromolecules 30:5290
102. Tomasi G, Scandola M, Briese BH, Jendrossek D (1996) Macromolecules 20:507
103. Spyros A, Kimmick R, Briese BH, Jendrossek D (1997) Macromolecules 30:8218
104. Mukai K, Yamada K, Doi Y (1993) Polym Degrad Stab 41:85
105. Kimura M, Toyota K, Iwatsuki M, Sawada H (1994) In: Doi Y, Fukuda K (), Biodegradable plastics and polymers. Elsevier Science, Amsterdam, p 92
106. Mergaert J, Andersson C, Wouters A, Swings J (1994) J Environ Polym Degrad 2:177
107. Yue CL, Gross RA, McCarthy SP (1996) Polym Degrad Stab 51:205
108. Doi Y, Kanesawa Y, Tanahashi N, Kumagai Y (1992) Polym Degrad Stab 36:173
109. Gilmore DF, Antoun S, Lenz RW, Fuller RC (1993) J Environ Polym Degrad 1:269
110. Eldsäter C, Albertsson A-C, Karlsson S (1997) Acta Polymerica 48:478
111. Eldsäter C, Albertsson A-C, Karlsson S (1999) Polym Degrad Stab 64:177
112. Hakkarainen M, Albertsson A-C, Karlsson S (1996) J Chromatogr A 741:251
113. Hakkarainen M, Albertsson A-C, Karlsson S (1997) J Appl Polym Sci 66:959

114. Karlsson S, Hakkarainen M, Albertsson A-C (1997) Macromolecules 30:7721
115. Karlsson S, Hakkarainen M, Albertsson A-C (1994) J Chromatogr A 688:251
116. Hennion M-C, Pichon V (1994) Environ Sci Technol 28:576A
117. Penton ZE (1997) In: Brown PR, Grushka E (eds), Advances in Chromatography. Marcel Dekker, New York, vol 37, chap 5
118. Kolb B (1982) Chromatographia 15:587
119. Hagman A, Jacobsson S (1987) J Chromatogr 395:271
120. Ramkumar D, Vaidya UR, Bhattacharya M, Hakkarainen M, Albertsson A-C, Karlsson S (1996) Eur Polym J 32:999
121. Karlsson S, Sares C, Renstad R, Albertsson A-C (1994) J Chromatogr A 669:97
122. Boyd-Boland AA, Chai M, Luo YZ, Zhang Z, Yang MJ, Pawliszyn JB, Gorecki T (1994) Environ Sci Technol 28:569A
123. Hakkarainen M, Albertsson A-C, Karlsson S (1997) J Environ Polym Degrad 5:67
124. Khabbaz F, Albertsson A-C, Karlsson S (1998) Polym Degrad Stab 61:329
125. Khabbaz F, Albertsson A-C, Karlsson S (1999) Polym Degrad Stab 63:127
126. Belbella A, Vauthier C, Fessi H, Devissaguet J-P, Puisieux F (1996) Int J Pharmac 129:95
127. Sbarbati del Guerra R, Cristallini C, Rizzi N, Barsacchi R, Guerra GD, Tricoli, Cerrai P (1994) J Mater Sci Mater Med 5:891
128. Karjomaa S, Suortti T, Lempiäinen R, Selin J-F, Itävaara M (1998) Polym Degrad Stab 59:333
129. Khabbaz F, Karlsson S, Albertsson A-C (2000) J Appl Polym Sci 78:2369
130. Kamei S, Inoue Y, Okada H, Yamada M, Ogawa Y, Toguchi H (1992) Biomaterials 13:953
131. Barrera DA, Zylstra E, Lansbury PT, Langer R (1995) Macromolecules 28:425
132. von Recum HA, Cleek RL, Eskin SG, Mikos AG (1995) Biomaterials 16:441
133. Vidil C, Braud C, Garreau H, Vert M (1995) J Chromatogr A 711:323
134. Braud C, Devarieux R, Garreau H, Vert M (1996) J Environ Polym Degrad 4:135
135. Braud C, Devarieux R, Atlan A, Ducos C, Vert M (1998) J Chromatogr B 706:73
136. Selembaron J, Marcincinova-Benabdillah K, Braud C, Vert M (2000) Polym Degrad Stab 68:281
137. Sanchez JG, Tsuchii A, Tokiwa Y (2000) Biotechnology Lett 22:849
138. Focarete ML, Ceccorulli G, Scandola M, Kowalczuk M (1998) Macromolecules 31:8485
139. Woo GLY, Mittelman MW, Santerre JP (2000) Biomaterials 2:1235
140. Braunegg G, Sonnleitner B, Lafferty RM (1978) Eur J Appl Microbiol Biotechnol 6:29
141. Karlsson S, Sares C, Albertsson A-C (1993) In: Schlegel HG, Steinbuchel A (eds), International Symposium on Bacterial Polyhydroxyalkanoates-ISBB'92. Goltze-Druck, Göttingen, p 455
142. Scandola M, Focarete ML, Adamus G, Sikorska W, Baranowska I, Swierczek S, Gnatowski M, Kowalczuk M, Jedlinski Z (1997) Macromolecules 30:2568
143. McLellan DW, Hallings PJ (1988) J Chromatogr 445:251
144. Athlan A, Braud C, Vert M (1997) J Environ Polym Degrad 5:243
145. Tokiwa Y, Iwamoto A, Koyama M, Kataoka N, Nishida H (1992) Makromol Chem Makromol Symp 57:273
146. Tanio I, Fukui T, Shirakura Y, Saito T, Tomita K, Kaiho T, Masamune S (1982) Eur J Biochem 124:71
147. Kasuya K, Doi Y, Yao T (1994) Polym Degrad Stab 45:379
148. Kasuya K, Inoue Y, Yamada K, Doi Y (1995) Polym Degrad Stab 48:167
149. Doi Y, Kitamura S, Abe H (1995) Macromolecules 28:4822
150. Grassie N, Murray EJ, Holmes PA (1984) Polym Degrad Stab 6:47
151. Renstad R, Karlsson S, Albertsson A-C (1997) Polym Degrad Stab 57:331
152. Renstad R (1998) PhD Thesis, KTH, Sweden
153. Weiler W, Gogolewski S (1996) Biomaterials 17:529
154. Migliaresi C, Cohn D, De Lollis A, Fambri L (1991) J Appl Polym Sci 43:83
155. Pegoretti A, Fambri L, Migliaresi C (1997) J Appl Polym Sci 64:213

156. Gogolewski S, Jovanovic M, Perren SM, Dillon JG, Hughes MK (1993) J Biomed Mater Res 27:1135
157. Vert M, Li S, Garreau H, Mauduit J, Boustta M, Schwach G, Engel R, Coudane J (1997) Die Angew Makromol Chem 247:239
158. Labreque LV, Kumar RA, Dave V, Gross RA, McCarthy SP (1997) J Appl Polym Sci 66:1507
159. Mochizuki H, Hirano M, Kanmuri Y, Kudo K, Tokiwa Y (1995) J Appl Polym Sci 55:289
160. Lee SH, Lee KH, Hong SK (1997) J Appl Polym Sci 64:1999
161. Jang SP, Lee KH, Kim MN (1997) Polymers for advanced Technologies 8:146
162. Grassie N, Murray EJ, Holmes PA (1984) Polym Degrad Stab 6:95
163. Verhoogt H, Ramsay BA, Favis BA, Ramsay JA (1996) J Appl Polym Sci 61:87
164. Hoffman A, Kreutzberger S, Hinrichsen G (1994) Polym Bull 33:355
165. Renstad R, Karlsson S, Albertsson A-C, Werner P-E, Westdahl M (1997) Polym Int 43:201
166. Renstad R, Karlsson S, Albertsson A-C (1999) Polym Degrad Stab 63:201
167. Parikh M, Gross RA, McCarthy SP (1993) In: Ching C, Kaplan DL, Thomas EL (eds), Biodegradable Polymers and Packaging. Technomic, Lancaster, p 159

Received: January 2001

Polymers from Renewable Resources

Margaretha Söderqvist Lindblad, Yan Liu, Ann-Christine Albertsson,
Elisabetta Ranucci, Sigbritt Karlsson

Department of Polymer Technology, Royal Institute of Technology,
100 44 Stockholm, Sweden
e-mail: aila @polymer.kth.se

Abstract. From the point of view of making novel polymers with inherent environment-favorable properties such as renewability and degradability, a series of interesting monomers are found in the metabolisms and cycles of nature. This review presents and discusses a number of aliphatic polyesters which show interesting applications as biomedical materials and degradable packages. Available from nature are amino acids, microbial metabolites from the conversion of glucose and other monosaccharides (e.g., acetic acid, acetone, 2,3-butanediol, butyric acid, isopropanol, propionic acid), lactic acid, ethanol and fatty acids. A series of biodegradable polymers with different properties and different potential industrial uses were made starting with succinic acid and/or 1,3-propanediol. There were two routes for making the polyester-based materials; the direct ring-opening polymerization of lactones (cyclic esters) synthesized from 1,3-propanediol, and the chain-extension of α,ω-dihydroxy-terminated oligomeric polyesters produced by thermal polycondensation of 1,3-propanediol and succinic acid (oligo(propylene succinate)s).

Keywords. Monomers from renewable resources, Polymers from renewable resources, 1,3-Propanediol, Succinic acid, Lactones, Cyclohexanedimethanol, Polyethyleneglycol, Chain-extension, Poly(ester-urethane)s, Poly(ester-carbonate)s

1	Introduction	141
2	Biosynthetic Sources for Monomers with a Potential for Making Novel Renewable Polymers	141
2.1	Amino Acids	143
2.2	Microbial Metabolites from the Conversion of Glucose and Other Monosaccharides	144
2.3	Homo- and Heterofermentation to Produce Lactic Acid and Ethanol	146
2.4	Fatty Acids	147
3	Synthetic Strategies to Develop Aliphatic Polyesters Deriving from Renewable Resources	148
3.1	Polyesters from Lactones Deriving from 1,3-Propanediol	149
3.2	Chain-Extension of α,ω-Dihydroxy-Terminated Oligomers	150

3.2.1 Synthesis of Oligo(1,3-Propylene Succinate)s by Thermal
       Polycondensation 151
3.2.2 Poly(Ester-Urethane)s 152
3.2.2.1 Chain-Extension Reaction by Diisocyanate Synthesis 152
3.2.2.2 Polymer Characterization 153
3.2.3 Poly(Ester-Carbonate)s 155
3.2.3.1 Chain-Extension Reaction of α,ω-Dihydroxy-Terminated Oligo
        (1,3-Propylene Succinate) 155
3.2.3.2 Chain-Extension of α,ω-Dihydroxy-Terminated Oligo
        (1,3-Propylene-Co-1,4-Cyclohexanedimethylene Succinate) 157
3.2.3.3 Segmented Copolymers Between Oligo(1,3-Propylene
        Succinate)s and Poly(Ethylene Glycol) 159

4   Conclusions 160

References 161

Abbreviations

ADP	Adenosine diphosphate
ATP	Adenosine triphosphate
DSC	Differential scanning calorimetry
DMA	Dynamic mechanical analyzer
DMAP	4-dimethylaminopyridine
EDIPA	N-ethyl-N,N-diisopropylamine
MDI	4,4′-diisophenylmethane diisocyanate
M_n	Number average molecular weight
M_w	Weight average molecular weight
NMR	Nuclear magnetic resonance spectroscopy
PD	Polydispersity
PEG	Poly(ethylene glycol)
PEU	Poly(ester-urethane)s
PHB	Poly-β-hydroxybutyrate
ROP	Ring-opening polymerization
SEC	Size exclusion chromatography
SP	Oligo(1,3-propylene succinate)
SP-C	Chain-extended product from SP using phosgene as chain-extending agent
SPC	Oligo(1,3-propylene-*co*-1,4-cyclohexanedimethylene succinate)
SPC-C	Chain-extended product from SPC using phosgene as chain-extending agent
T_g	Glass transition temperature
T_m	Melting point

1
Introduction

Industrial biotechnology has evolved as a significant manufacturing tool for products such as ethanol, organic acids and bulk amino acids as speciality products for food and pharmaceutical applications [1]. Polyesters represent one of the most promising families of biodegradable polymers [2–6], whose potential applications cover such widely different fields as packaging for industrial products, mulching for agriculture, bioresorbable biomaterials for hard tissue replacement and controlled drug delivery systems. The potential to create a range of polyesters with varying properties from hard to soft and elastic by blending, co-polymerization or the use of natural additives offers the opportunity to create highly sophisticated materials for use as e.g., biomaterials as well as simpler, less expensive plastics for use in packages. The degradation behavior of such materials may be tailored so that adequate degradation time is achieved. At the same time, the number of degradation products is relatively small and this means that toxicity testing is relatively easy. The degradation products generally fit well into one of the numerous metabolism processes occurring in nature. A prerequisite for making a "green" polymer is a cost-effective production of the monomer; a good substrate in that sense is glucose [7]. An estimate of the product costs for a 10 000 ton per annum plant indicates that the microbial process can become more attractive than the chemical route.

This paper is a summary of new biodegradable polymers from renewable resources.

2
Biosynthetic Sources for Monomers with a Potential for Making Novel Renewable Polymers

A range of chemically interesting compounds for use in making new polymers is produced by the natural metabolic processes. Some of these compounds are true end products, while others are intermediates in different cycles. Some are bacterial or fungal metabolites while yet others may be isolated from plants or wood. In principle, we may use all available α-amino acids, monosaccharides, the intermediates of the citric acid cycle, alcohol- and acid-fermentation products, and also different carboxylic acids such as those from soybean or rape (palmitic, stearic, oleic, linoleic and linolenic acid). The major bio-elements are assimilated in an inorganic form; carbon as CO_2, nitrogen as ammonia, nitrate or N_2, sulphur as sulphate, and hydrogen and oxygen as water. The precursors of macromolecules are not only amino acids, monosaccharides, nucleic acids and lipids, but also sugar phosphates, pyruvates, acetates, oxalacetates, succinic acid and α-ketoglutarate, formed during catabolism by heterotrophs or in CO_2 assimilation by autotrophs. A class of modern plastics from renewable resources are available by exploitation of these compounds as monomers. Scheme 1 presents the use of aromatic compounds for respiration by aerobic bacteria.

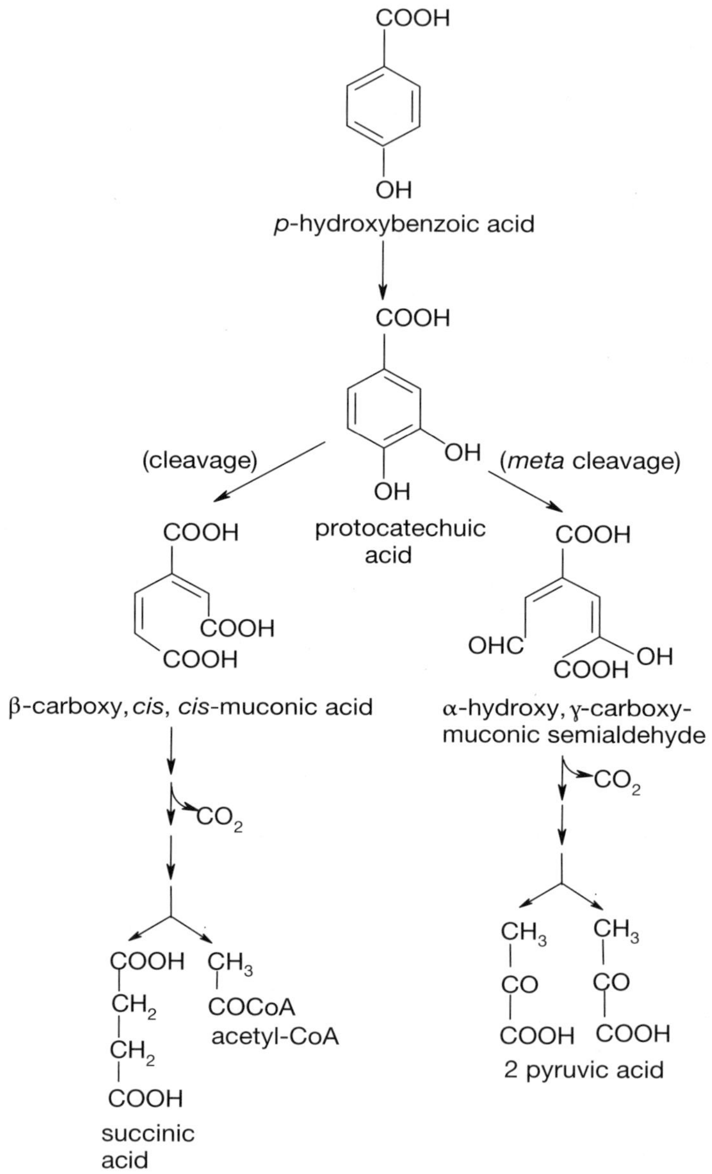

Scheme 1. The divergent pathways for the oxidation of *p*-hydroxybenzoate for formation of e.g., succinic acid

This β-ketoadipate pathway produces e.g., succinic acid, a compound used as monomer for the production of aliphatic polyesters. Already in 1881, the production of e.g., 1,3-propanediol by the fermentation of glycerol was reported [8].

2.1
Amino Acids

There are different routes to produce α-amino acids. Three routes assimilate nitrogen by the use of ammonia, which makes oxidation or reduction unnecessary. The amino acids, glutamic acid, asparagine and glutamine are formed from α-ketoglutaric acid, aspartic acid and glutamic acid respectively. Both glutamic acid and glutamine serve to transfer amino groups to other nitrogenous precursors of cellular macromolecules. In general, the syntheses of amino acids are grouped in five "families"; the glutamate, aspartate, aromatic, serine and pyruvate types. Table 1 lists the precursors and the resulting amino acids. The pathways use different enzymes and coenzymes. Scheme 2 shows in more detail how e.g., intermediates from the citric acid cycle are converted to amino acids. Some of the amino acid biosynthesis also leads to the formation of intermediates which are converted to essential cell materials such as folic acid, *p*-hydroxybenzoic acid, diaminopimelic acid, dipicolinic acid and purines. The metabolism of a micro-organism is precisely regulated, which means that the organism normally synthesizes the quantities of amino acids sufficient to meet its require-

Table 1. Biosynthetic derivations of amino acids

Biosynthetic Derivations of Amino Acids

Precursor Metabolite		Amino Acid	Family
Oxaloacetate	⟶ Aspartate	Asparagine Methionine Threonine ↓ Isoleucine Lysine	Aspartate
α-Ketoglutarate	⟶ Glutamate	Glutamine Arginine Proline	Glutamate
Phosphoenolpyruvate + erythrose-4-phosphate	⟶	Tryptophan Phenylalanine Tyrosine	Aromatic
β-Phosphoglycerate	⟶ Serine	Glycine Cysteine	Serine
Pyruvate	⟶	Alanine Valine Leucine	Pyruvate
Ribose-5- phosphate + ATP	⟶	Histidine	

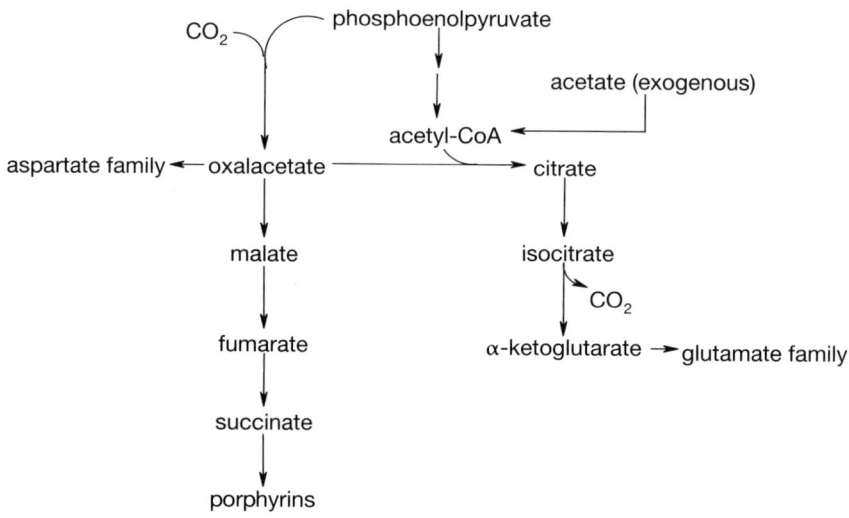

Scheme 2. Biosynthesis of some amino acids

ment. Natural and mutant strains of some micro-organisms have defective mechanisms for the regulation of their biosynthesis and the production of large amounts of a specific amino acid is thus possible. Genetically modified species are probably sources of specific amino acids, which serve as monomers for the production of synthetic proteins.

2.2
Microbial Metabolites from the Conversion of Glucose and Other Monosaccharides

The conversion of sugars to pyretic acid is the initial step towards the production of many interesting monomers such as acetic acid, acetone, 2,3-butanediol, butanol, butyric acid, isopropanol and propionic acid. The Embden-Meyerhof (glycolytic pathway), the Hexose Monophosphate Shunt and the Entner-Duodoroff reactions produce the key intermediates. Besides the fermentation of pyruvate to the above monomers, pyruvate is also oxidized to produce adenosine triphosphate (ATP) by the citric acid cycle (or *Tricarboxylic Acid Cycle*). A special modification of the citric acid cycle is the glyoxalate cycle for the oxidation of e.g., acetic acid or higher fatty acids to acetyl-S-CoA. *Iso*citric acid, a normal intermediate of the citric acid cycle is cleaved by two reactions to yield succinic acid and glyoxylic acid, and acetyl-*S*-CoA is subsequently condensed with glyoxylic acid to yield malic acid. Both succinic and malic acids are monomers for the production of degradable aliphatic polyesters.

Scheme 3 shows the end products of bacterial fermentation and the pathways by which they are formed. Most bacterial fermentations produce several end

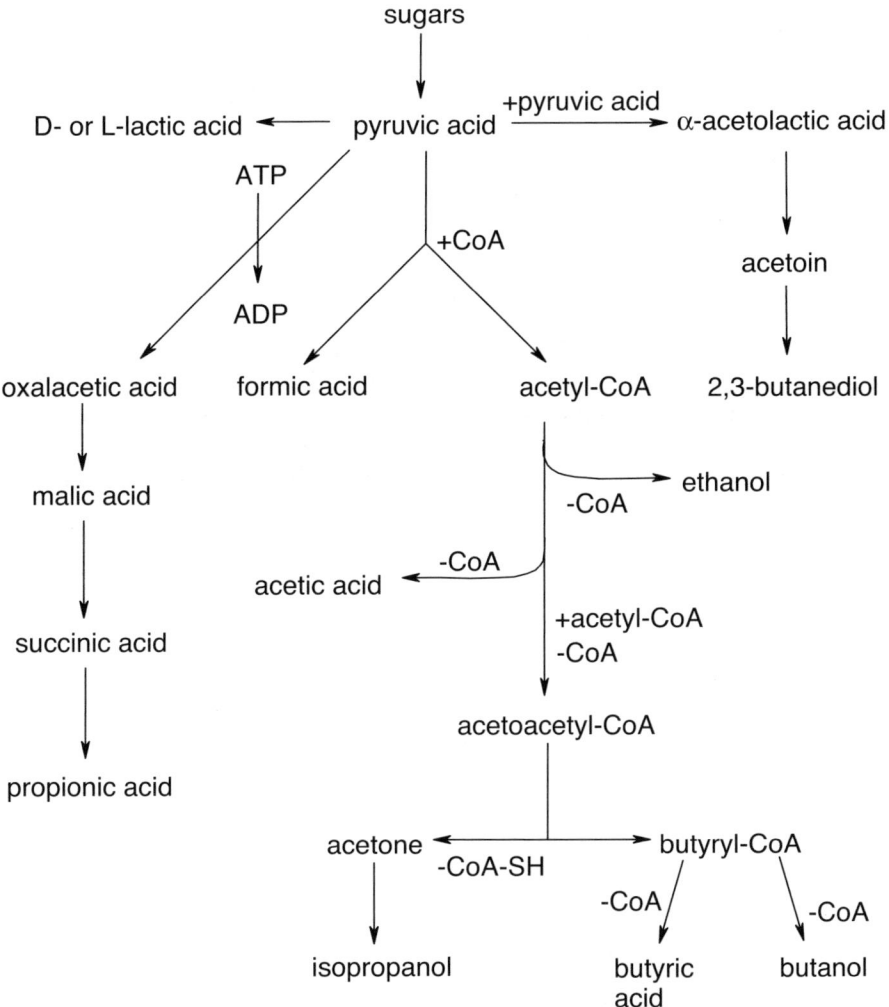

Scheme 3. Bacterial fermentation of sugars and production of possible monomers

products, but all organisms do not produce all the products in the scheme. The end products formed are in most cases group-specific.

Species of *Clostridium* have been used on a very large scale for the production of industrial solvents such as acetone and butanol. Many clostridia ferment sugars with the formation of carbon dioxide, hydrogen and butyric acid. Some of them convert butyric acid to butanol and the acetic acid to ethanol and acetone. The acetone-butanol process by *Clostridium acetobutylicum* expanded after its introduction just before World War I. Although the industry has now been almost totally replaced by the synthetic production from oil, it may again be of interest for the production of renewable monomers.

2.3
Homo- and Heterofermentation to Produce Lactic Acid and Ethanol

Lactic acid bacteria produce large amounts of lactic acid from glucose or other sugars. Homo- and heterofermentative bacteria, e.g., *Streptococcus, Leuconostoc* and *Lactobacillus,* ferment lactic acid. The homofermenters convert glucose almost quantitatively to lactic acid while the heterofermenters produce an equimolar mixture of lactic acid, ethanol and CO_2. Scheme 4 shows the production of lactic acid from glucose. The lactic acid bacteria produce isomers of lactic acid and the stereoform of the isomers produced is determined by the stereospecificity of the lactic dehydrogenase, which mediates the pyruvate reduction. Some species contain only D-lactic dehydrogenase and from the D-isomer while others contain only the L-lactic dehydrogenase and from the L-isomer. Certain species have two hydrogenases of differing stereospecificity and hence form racemic lactic acid. Scheme 4 also shows the regulation in lactic acids, a regulation that inhibits or activates the production of lactate dehydrogenase and pyruvate formate-lyase which produces ethanol and acetic acid. In the production of lac-

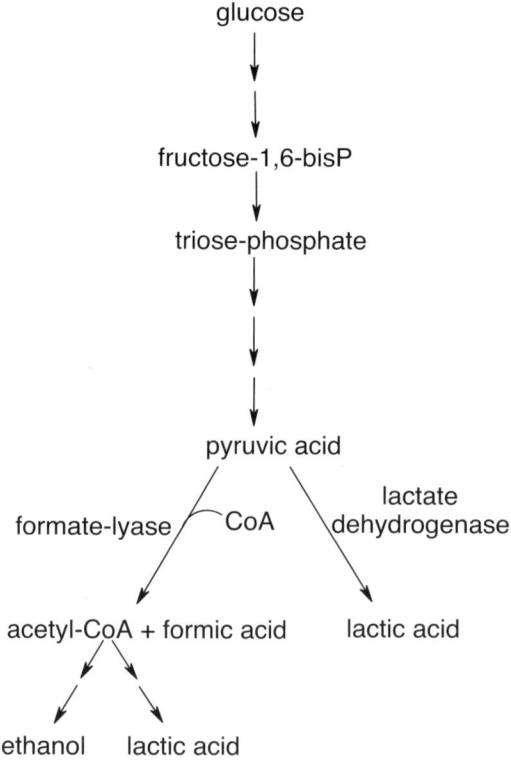

Scheme 4. Hetero- and homofermentation from glucose to lactic acid and ethanol

tic acids, the regulation of the pyruvate metabolism is achieved by glycolytic intermediates such as fructose-1,6-bis-phosphate or triose-phosphate.

2.4
Fatty Acids

The lipids are chemically heterogeneous and consist of fats, phospholipids, steroids, isoprenoides and poly-β-hydroxybutyrate (PHB). These are in principle either lipid-containing esterified fatty acids or those consisting of repeating C_5 units (isoprene). Esterified fatty acids in bacteria may be straight-chained or branched; some have double bonds and/or hydroxyl groups. However, most bacterial fatty acids contain an even number of carbon atoms. Monosaturated fatty acids are formed in many bacteria by one of two routes; the aerobic pathway or the anaerobic pathway. The aerobic route introduces the unsaturated from the fully saturated fatty acid, while the anaerobic instead occurs during elongation of the fatty acid chain.

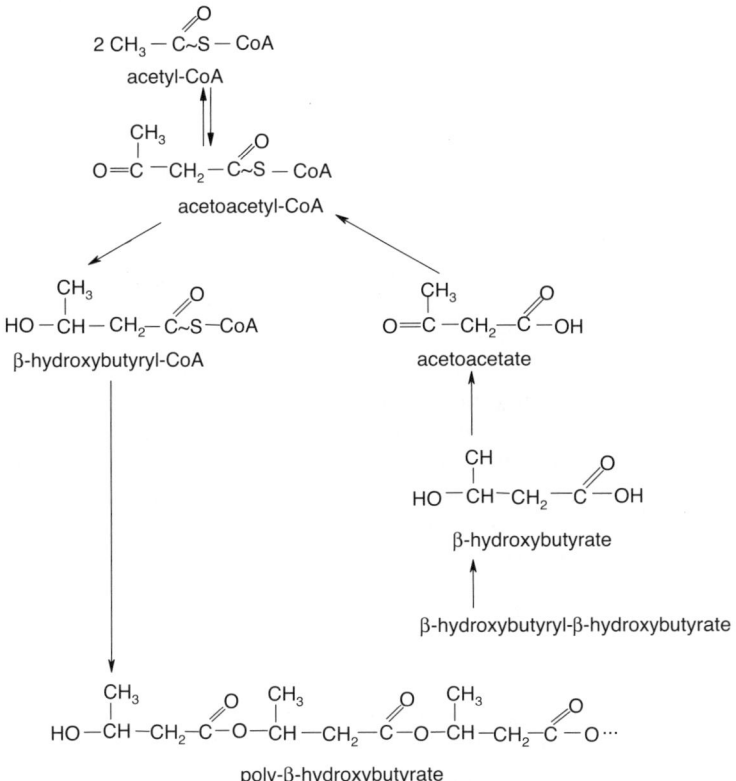

Scheme 5. The reactions involved in the synthesis and degradation of poly-β-hydroxybutyrate

PHB is organic reserve material in several bacteria such as *Pseudomonas, Azotobacter, Rhizobium, Bacillus* etc. while glycogen is stored in many spore-forming *Bacillus* and *Clostridium*. The synthesis of PHB is a way of accumulating a carbon store in a form that is osmotically inert and a way of neutralizing an acidic metabolite. Scheme 5 shows the synthesis and degradation of PHB. PHB is synthesized as a side path on the metabolic route of fatty acid synthesis. Various derivatives of PHB have been presented where the carbon source is shifted from glucose to various others, e.g., propionic or octanoic acid, producing polyhydroxyalkanoates with better polymeric properties than the native PHB.

Soybean and rape oils from plants, e.g palmitic, stearic, oleic, linoleic and linolenic acid, are also available for the production of new polymeric materials.

3
Synthetic Strategies to Develop Aliphatic Polyesters Deriving from Renewable Resources

1,3-propanediol and succinic acid can both be obtained by the fermentation of glycerol produced by cleavage of natural fats, using e.g., *Clostridia, Enterobacteriaceae* or *Klebsiella* [7, 9]. A more cost-effective production of 1,3-propanediol than from glycerol, which can be achieved by using glucose, is required. Great efforts are therefore now being made to combine the pathway from glucose to glycerol with the bacterial route from glycerol to 1,3-propanediol [7]. Yields of succinic acid as high as 110 g/l have been achieved from glucose using newly discovered rumen organisms (*Actinobacillus succinogenes*) [10, 11].

Polyester-based materials have basically been prepared following two main synthetic strategies:
– The first group of high molecular weight polyesters were produced by the direct ring-opening polymerization (ROP) of lactones (cyclic esters) synthesized from 1,3-propanediol.
– The second group were produced by the chain-extension reaction of α,ω-dihydroxy-terminated oligomeric polyesters produced by the thermal polycondensation of 1,3-propanediol and succinic acid (oligo(propylene succinate)s). Depending on the chain-extension technology adopted, poly(ester-urethane)s or poly(ester-carbonate)s were obtained. In the case of poly(ester-carbonate)s, the chain-extended products of α,ω-dihydroxy-terminated oligomeric copolyesters were also produced using a different alcoholic moiety, i.e. 1,4-cyclohexanedimethanol. Segmented copolymers of α,ω-dihydroxy-terminated oligo(propylene succinate)s and poly(ethylene glycol) (PEG) were also synthesized.y

All the materials produced were characterized in terms of M_n, M_w and PD by SEC carried out using dimethylformamide as mobile phase. Their structures were also identified by NMR spectroscopy. All characterizations were in agreement with the proposed structures. Their main thermal parameters were characterized by DSC. In some cases, preliminary processing trials were carried out.

Mechanical (Instron testing) and dynamic mechanical (DMA) tests were also performed in some cases.

3.1
Polyesters from Lactones Deriving from 1,3-Propanediol

The production of polyesters from lactones (cyclic esters) avoids the problems usually encountered in traditional polyester synthesis due to esterification equilibrium and water production. In fact, they yield in a single step, in the presence of ROP catalysts, high molecular weight polymers endowed with optimum mechanical properties. This is particularly true when large lactones are used as starting monomers. On the other hand, one of the main drawbacks connected with their use lies in their high cost.

We have spent much effort in establishing methods for synthesizing a specific lactone deriving from 1,3-propanediol, i.e., 1,4-dioxepan-2-one and for polymerizing it according to standard ROP technologies (see Scheme 6).

1,4-Dioxepan-2-one was synthesized according to a strategy implying the reaction of 1,3-propanediol with chloroacetic acid sodium salt (see Scheme 7). During the first stage, 1,3-propanediol was first transformed to its sodium mono-alcoholate, and then coupled with chloroacetic acid sodium salt and finally thermally self-condensed to the corresponding polymer. During the final stage, which was a vacuum fractional distillation, the dimer of 1,4-dioxepan-2-one was obtained (1,4,8,11-tetraoxacyclotetradecane-2,9-dione) via the thermal decomposition of the above polymer under vacuum. It can be polymerized by ROP and, the polymerization steps being the same, it yields much higher molecular weights than the polymerization of the corresponding monomer.

The polymerization of the cyclic dimer (1,4,8,11-tetraoxacyclotetradecane-2,9-dione) was carried out according to traditional ROP procedures using $SnOCt_2$ as catalyst (see Scheme 8). The reaction was performed in the bulk, under an inert atmosphere, at 150 °C. Different reaction times, different temperatures and different catalyst concentrations were used and the results were compared. The molecular weight varied in the range 20,000–80,000, depending on the synthesis conditions adopted.

The whole procedure, although valuable for producing high molecular weight polymers, has a weak point in the low yields obtained in the monomer synthesis. Although several attempts were made to improve the synthetic pathway, yields were never higher than 30–40%.

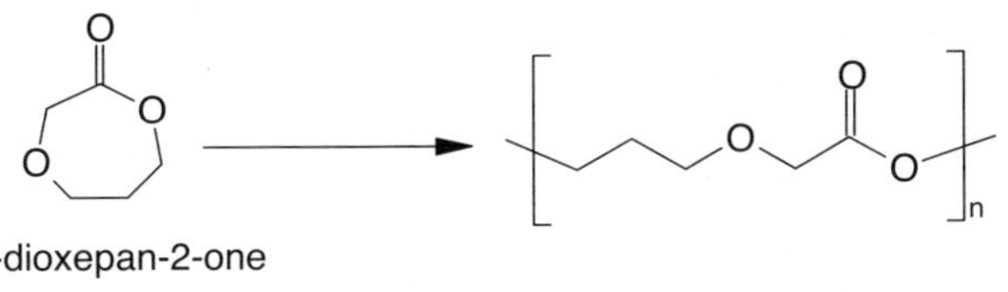

1,4-dioxepan-2-one

Scheme 6. ROP of 1,4-dioxepan-2-one

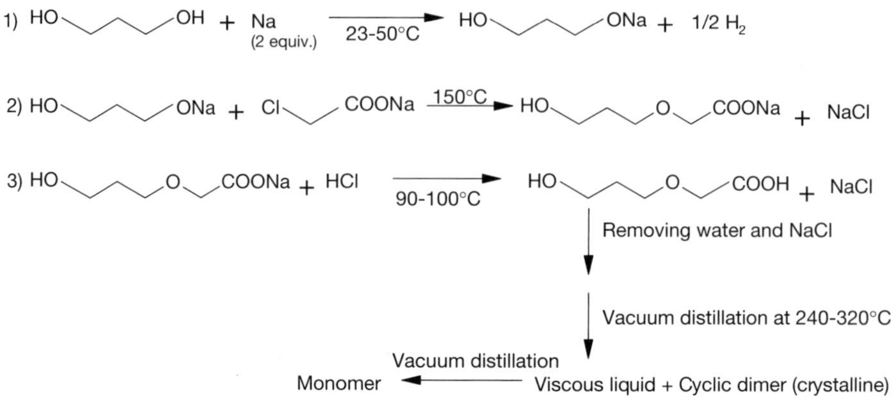

1) HO~~~OH + Na (2 equiv.) →(23-50°C) HO~~~ONa + 1/2 H₂

2) HO~~~ONa + Cl~COONa →(150°C) HO~~~O~COONa + NaCl

3) HO~~~O~COONa + HCl →(90-100°C) HO~~~O~COOH + NaCl
   ↓ Removing water and NaCl
   ↓ Vacuum distillation at 240-320°C
   Monomer ←(Vacuum distillation) Viscous liquid + Cyclic dimer (crystalline)

Scheme 7. Synthesis of 1,4-dioxepan-2-one from 1,3-propanediol and chloroacetic acid sodium salt

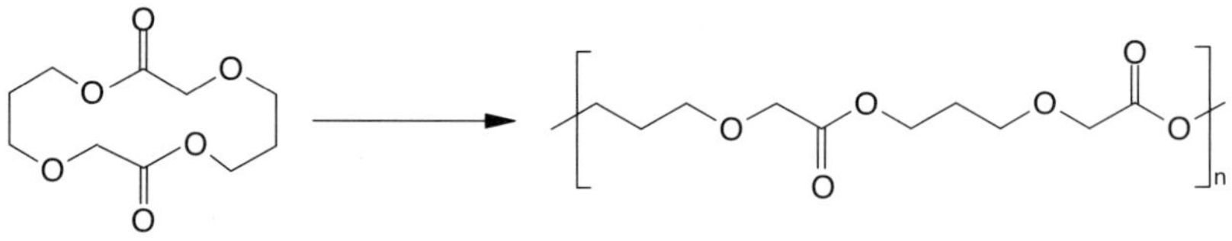

1,4,8,11-tetraoxacyclotetradecane-2,9-dione

Scheme 8. ROP of 1,4,8,11-tetraoxacyclotetradecane-2,9-dione

3.2
Chain-Extension of α,ω-Dihydroxy-Terminated Oligomers

Polyester synthesis via diacid/dialcohol thermal polycondensation is the traditional and most economic procedure for obtaining polyesters. Unfortunately, it encounters problems due to the difficulty in obtaining high molecular weight materials, due to a slow equilibrium and to hydrolysis in the presence of water obtained as side product. A useful way to overcome the problems of incomplete polymerization and short chain length is by the so-called chain–extension technology, usually carried out in the presence of coupling agents reacting with the reactive functions located at prepolymer chain ends. It is well known, in fact, that an increase in the molecular weight of a polymeric material means an improvement in the physical performance of the polymers.

Two different types of chain-extending agents were used, namely 4,4′-diisophenylmethane diisocyanate (MDI) and phosgene, which lead to two different classes of copolymers, poly(ester-urethane)s and poly(ester-carbonate)s respectively. High molecular weight poly(ester-carbonate)s have earlier been prepared by the dichloroformate synthesis [12–13].

One way of improving the physical and mechanical properties of the polymeric materials produced was to copolymerize the original monomers with stiffer comonomers, whose corresponding homopolymers are endowed with high thermal and mechanical properties. Therefore, 1,4-cyclohexanedimethanol was copolymerized with 1,3-propanediol and succinic acid. Segmented copolymers between oligo(propylene succinate)s and PEG were also produced in order to modify their hydrophilic/hydrophobic properties.

3.2.1
Synthesis of Oligo(1,3-Propylene Succinate)s by Thermal Polycondensation

Much effort has been devoted to the optimization of the polyesterification reaction. For instance, different types of monomeric precursors structurally related to succinic acid (e.g., dimethyl succinate or succinic anhydride) were used. Different kinds of catalysts (e.g., phenolates, titanium alkoxides, tin octanoates) at different concentrations were studied. Different reaction temperatures (130–190 °C) were reached and different procedures for water elimination (vacuum drying under different conditions or toluene distillation) were adopted. Experimental results obtained showed that the use of different catalysts and different monomer precursors (succinic acid derivatives) did not significantly alter the polymerization kinetics or yield, and for this reason, they were abandoned. The procedure finally adopted is summarized below.

Different α,ω-dihydroxy-terminated oligo(propylene succinate)s (SP) were prepared by the thermal polycondensation of excess 1,3-propanediol and succinic acid, as illustrated in the first part of Scheme 9 [4]. The molar ratio of 1,3-propanediol to succinic acid was maintained in the range of 1.05–1.25. The molecular weight of the products depended on the excess 1,3-propanediol used, but it usually ranged between 900 and 2400. The molecular characterization and appearance of some typical samples are given in Table 2.

Table 2. Molecular weights of oligo(propylene succinate)s

Sample	OH/COOH Molar ratio	$\overline{M_n}$	$\overline{M_w}$	PD[a]	Appearance
SP1	1.20	900	1200	1.3	Viscous liquid
SP2	1.20	1400	1800	1.3	Soft wax
SP3	1.10	2400	3500	1.5	Hard wax
SP4	1.15	1100	1500	1.3	Soft wax
SP5	1.05	2300	3400	1.5	Hard wax
SP6	1.25	1200	1600	1.4	Viscous liquid

[a] PD = polydispersity index

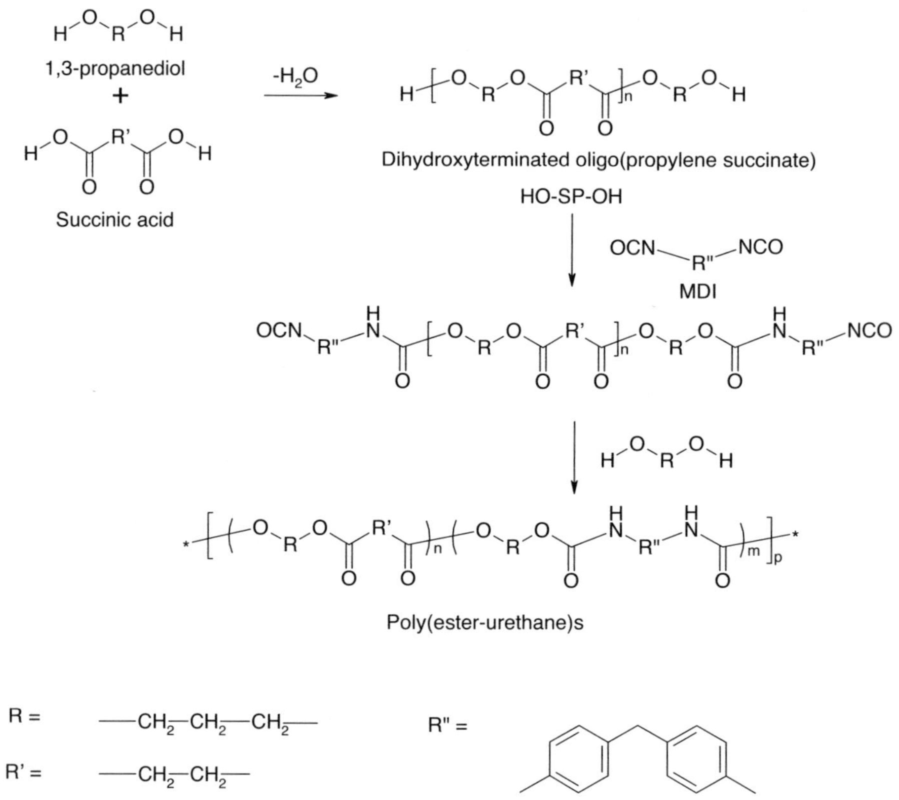

Scheme 9. Synthesis of oligo(propylene succinate), by thermal polycondensation of 1,3-propanediol and succinic acid, and chain-extension reaction by diisocyanate synthesis

3.2.2
Poly(ester-urethane)s

3.2.2.1
Chain-Extension Reaction by Diisocyanate Synthesis

High molecular weight poly(ester-urethane)s (PEU) were prepared *via* the diisocyanate synthesis, as illustrated in the second part of Scheme 9 [4]. Oligo(propylene succinate)s having molecular weights in the range of 2300–2400 were chain-extended in the presence of MDI, affording fairly high molecular weight PEU. A series of materials with different distributions of the polyester and polyurethane segments were produced and characterized. The most representative materials obtained are listed in Table 3.

Table 3. Molecular weights of poly(ester urethane)s

Sample	Prepolymer in feeding (weight %)	$\overline{M}_n$	$\overline{M}_w$	PD[a]
PEU 46[b]	46.0	34,400	64,000	1.9
PEU 49[c]	49.0	23,200	40,600	1.8
PEU 51[b]	51.2	24,000	41,400	1.7
PEU 52[c]	52.0	20,000	33,800	1.7
PEU 55[c]	55.0	25,200	46,400	1.8
PEU 56[b]	56.5	18,200	29,400	1.6
PEU 58[c]	58.0	20,300	34,200	1.7
PEU 60[b]	59.8	19,600	32,100	1.6
PEU 61[b]	61.4	13,600	22,300	1.6
PEU 63[c]	63.0	21,700	41,600	1.9

[a] PD = polydispersity index, [b] Synthesized from SP3, [c] Synthesized from SP5

3.2.2.2
Polymer Characterization

The series of ten different segmented poly(ester-urethane)s obtained was characterized by SEC (molecular weight, polydispersity), DMA (thermo-mechanical properties), DSC (thermal properties) and Instron (mechanical properties). Before the mechanical testing could be performed, it was necessary to establish procedures for film production.

The influence of the processing conditions during film production was studied with a statistical experimental design for two different PEU copolyesters. Temperature, delay time before pressure application and pressure application time were factors considered. The experiments were performed with a high value, a low value and a center value for each factor. Table 4 shows the worksheet relative to one of the PEU copolyesters. The selected low temperature value was the onset temperature of melting determined by DSC and the high temperature value was the melting point. The pressure was maintained constant and equal to 200 bars, and the films were cooled in air. Molecular weight, (M_n), and polydispersity were determined for all of the films (Table 4). The best processing conditions were low temperature, a short delay time before pressure application and a short pressure application time. Results for the whole series, obtained when using optimum processing conditions, are given in Table 5.

Table 6 reports the storage modulus in the rubbery region and T_g measured at maximum tan δ by DMA for PEU copolymers at different polyester contents. Standard deviations are also given. T_g values show fairly high standard deviations, probably because of the heterogeneity in the sample morphology. The storage modulus in the rubbery region decreased, as expected, with increasing content of aliphatic polyester. T_g varied from 4.8 °C to –7.8 °C and the peak values shifted to a somewhat lower temperature as the content of aliphatic polyester increased. The melting point (T_m) of PEU copolymers measured by DSC varied

Table 4. Worksheet for investigation of the influence of the conditions during the film production for PEU 46

Temperature, °C	Pressure application, min	Delay time before pressure application, min	$\overline{M}_n$	PD[a]
172	10	1	34,600	2.1
208	10	1	12,700	1.5
172	20	1	27,700	2.0
208	20	1	9,700	1.4
172	10	5	30,100	2.0
208	10	5	10,500	1.4
172	20	5	28,800	2.0
208	20	5	11,000	1.5
190	15	3	16,600	1.6
190	15	3	20,800	1.7
190	15	3	18,800	1.6

[a] Polydispersity index

Table 5. Molecular weights for PEU copolyesters as powders and as films

Sample	$\overline{M}_n$ (powder)	$\overline{M}_n$ (film)	PD[a] (powder)	PD[a] (film)
PEU 46	34,400	26,800	1.9	2.1
PEU 49	23,200	22,900	1.8	1.9
PEU 51	24,000	23,200	1.7	1.9
PEU 52	20,000	20,700	1.7	1.8
PEU 55	25,200	24,700	1.8	2.0
PEU 56	18,200	19,400	1.6	1.7
PEU 58	20,300	21,700	1.7	1.8
PEU 60	19,600	19,600	1.6	1.7
PEU 61	13,600	13,800	1.6	1.7
PEU 63	21,700	19,100	1.9	2.3

[a] Polydispersity index

Table 6. Results of dynamic mechanical measurements. Mean values and standard deviations from four measurements of each sample are reported

Sample	Prepolymer in feeding (weight %)	Storage modulus[a] (MPa)	T_g[b] (°C)
PEU 46[c]	46.0	364±19	4.8±1.9
PEU 49[d]	49.0	204±8	4.0±1.7
PEU 51[c]	51.2	171±13	0.8±2.0
PEU 52[d]	52.0	127±13	2.0±0.4
PEU 55[d]	55.0	106±2	−0.5±1.1
PEU 56[c]	56.5	77±3	−2.1±1.1
PEU 58[d]	58.0	64±9	−1.8±3.4
PEU 60[c]	59.8	47±6	−3.6±3.5
PEU 61[c]	61.4	36±5	−6.5±2.7
PEU 63[d]	63.0	33±1	−7.8±2.3

[a] Rubbery region; 23 °C, [b] Measured at tan δ, [c] Synthesized from SP3, [d] Synthesized from SP5

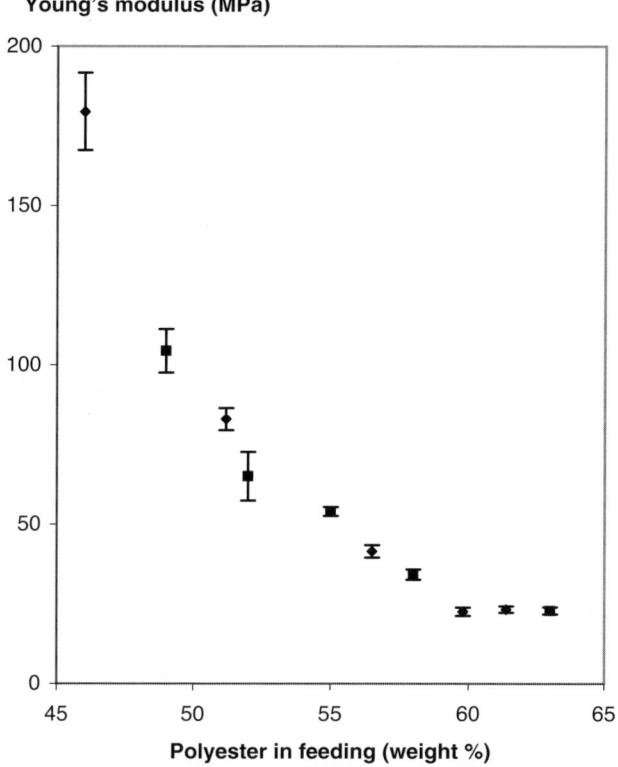

Fig. 1. Young's modulus for PEU copolymers versus feeding: (♦) PEU series from SP3 and (■) PEU series from SP5

in the range from 176 °C to 208 °C. The sample with the highest content of aliphatic polyester showed, as expected, the lowest melting point.

Figure 1 shows the Young's modulus as a function of polyester content in the feed. In agreement with earlier results, Young's modulus decreased when the content of aliphatic polyester increased.

3.2.3
Poly(ester-carbonate)s

3.2.3.1
Chain-Extension Reaction of α,ω-Dihydroxy-Terminated Oligo(1,3-Propylene Succinate)

High molecular weight poly(ester-carbonate)s were prepared *via* the dichloroformate synthesis as illustrated in Scheme 10 [3].

This process involves two steps. In the first step, α,ω-dihydroxy-terminated oligo(propylene succinate)s (SP) were prepared by the thermal polycondensation of excess diol and diacid, (see first part of Scheme 9). In the second step, the

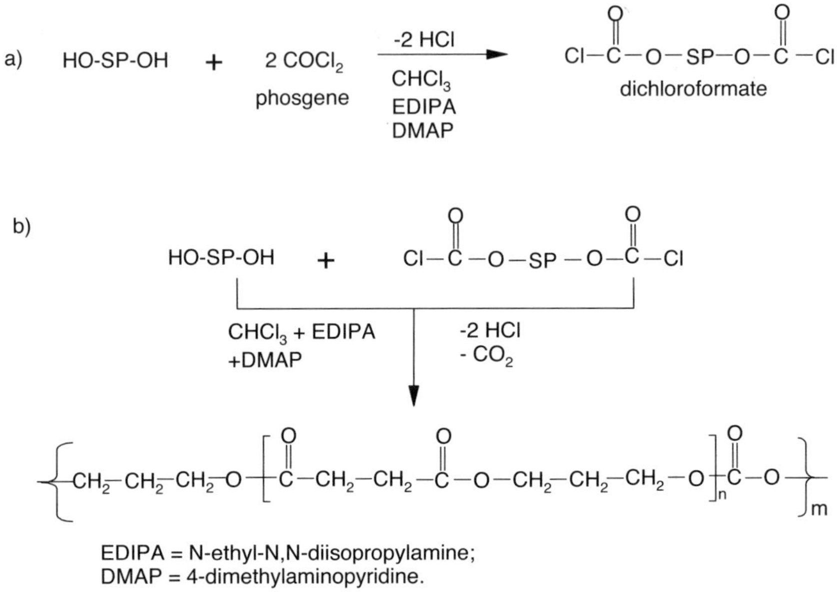

a) HO-SP-OH + 2 COCl$_2$ $\xrightarrow[\substack{CHCl_3 \\ EDIPA \\ DMAP}]{-2\ HCl}$ Cl–C(=O)–O–SP–O–C(=O)–Cl

phosgene   dichloroformate

b) HO-SP-OH + Cl–C(=O)–O–SP–O–C(=O)–Cl

        CHCl$_3$ + EDIPA   -2 HCl
        +DMAP   - CO$_2$

$\{-CH_2-CH_2-CH_2-O-[-C(=O)-CH_2-CH_2-C(=O)-O-CH_2-CH_2-CH_2-O-]_n-C(=O)-O-\}_m$

EDIPA = N-ethyl-N,N-diisopropylamine;
DMAP = 4-dimethylaminopyridine.

Scheme 10. Chain-extension reaction by dichloroformate synthesis of oligo(propylene succinate)

oligomers were chain-extended by using phosgene. The chain-extension reaction, in turn, was performed in two steps. The oligomers were dissolved in dry chloroform in the presence of acid acceptors (e.g., ethyldiisopropylamine). They were then treated with excess phosgene and the unreacted phosgene was eliminated by a stream of dry nitrogen. The resulting α,ω-dichloroformate was then polycondensed with an equivalent amount of α,ω-dihydroxy-terminated oligo(propylene succinate)s. The carbonate chain-extended oligo(propylene succinate) (SP-C) is essentially a high-molecular-weight polyester-based polymer containing some carbonate bonds along the macromolecular chain.

The thermal properties of SP-C, evaluated by means of DSC, showed that, even in fairly high molecular weight materials, the observed melting point was fairly low, 48 °C (see Table 7). It may be interesting to observe that the passage from the oligomer (SP) to its high molecular weight chain-extension product (SP-C) results in a 6.3 °C increase in the glass transition temperature and in only a modest increase in the crystalline melting point (1.2 °C). It is worth noticing, however, that the melting peak of the oligomer is bimodal, and the reported melting point corresponds to the higher melting peak. On the other hand the melting enthalpy decreases (−11 J/g), indicating a lower degree of crystallinity in SP-C. This may be because the chain-extension introduces structural irregularities along the polyester chain. The storage modulus in the rubbery region, measured by DMA, was only 188 MPa.

3.2.3.2
Chain-Extension of α,ω-Dihydroxy-Terminated Oligo(1,3-Propylene-Co-1,4-Cyclohexanedimethylene Succinate)

In order to improve the thermal and mechanical properties of the poly(ester-carbonate)s, chain-extension products (SPC2-C and SPC3-C) were also synthesized starting from dihydroxy-terminated oligomers obtained by the esterification of succinic acid in the presence of 1,3-propanediol and 1,4-cyclohexanedimethanol (SPC1, SPC2 and SPC3)[5]. Scheme 11 shows the synthesis of α,ω-dihydroxy-terminated oligo(1,3-propylene-co-1,4-cyclohexanedimethylene succinate) and Scheme 12 shows the chain-extension reaction. Molecular weights and molecular weight distributions of the polymers obtained, as well as the transition temperatures, are reported in Table 7.

Table 7. Molecular and thermal characterizations of poly(ester-carbonate)s

Sample	$\overline{M}_n$	$\overline{M}_w$	PD	T_g (°C)	T_m (°C)	ΔH (J/g)
SP Oligomer from 1.02 PrOH/1.00 SA (mole)	2200	2900	1.3	–35.4	46.8	50.4
SPC 1 Oligomer from PrOH (70%)/CHDM (30%)/SA (mole)	3100	4400	1.4	–26.3	47.8	5.7
SPC 2 Oligomer from PrOH (50%)/CHDM (50%)/SA (mole)	2600	3600	1.4	–21.8	47.8	23.0
SPC 3 Oligomer from PrOH (30%)/CHDM (70%)/SA (mole)	2900	4100	1.4	–15.2	84.0	30.8
SP-C Chain-extended polymer from oligomer SP	30,000	48,000	1.6	–29.1	48.0	39.4
SPC 2-C Chain-extended polymer from oligomer SPC 2	32,000	50,000	1.6	–16.8	48.7	12.9
SPC 3-C Chain-extended polymer from oligomer SPC 3	15,000	24,000	1.6	–10.4	76.1	30.9
SP-PEG 1000 Copolymer from PEG 1000 and SP	47,000	66,000	1.4	–42.5	35.3	44.2
SP-PEG 2000 Copolymer from PEG 2000 and SP	30,000	53,000	1.8	–43.5	46.7	61.1

PrOH: 1,3-propanediol; SA: succinic acid; CHDM: 1,4-cyclohexanedimethanol; PEG: poly(ethylene glycol)

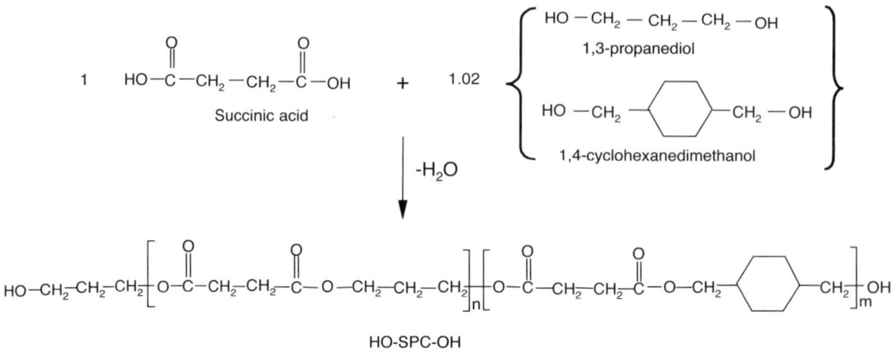

Scheme 11. Synthesis of oligo(1,3-propylene-*co*-1,4-cyclohexanedimethylene succinate)

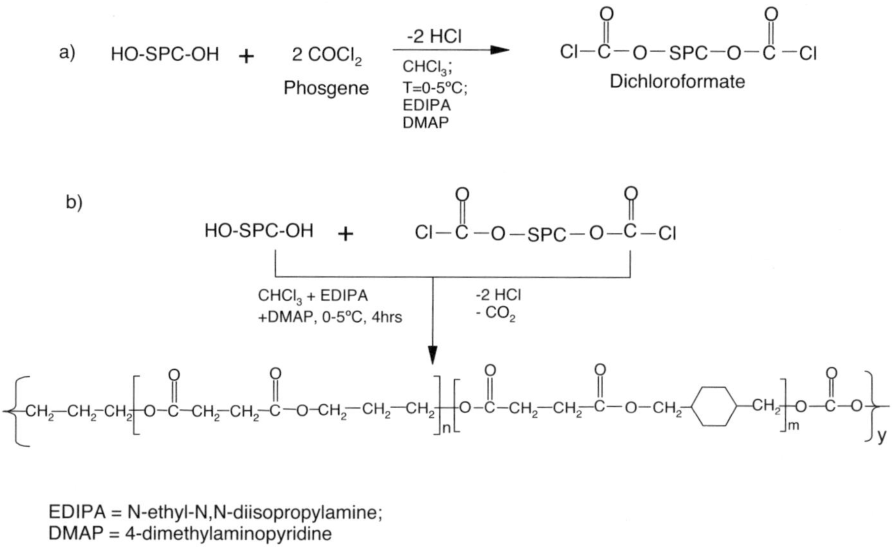

Scheme 12. Chain-extension reaction by dichloroformate synthesis of oligo(1,3-propylene-*co*-1,4-cyclohexanedimethylene succinate)

EDIPA = N-ethyl-N,N-diisopropylamine;
DMAP = 4-dimethylaminopyridine

The results of the thermal characterization can be summarized as follows:
- T_g increases with increasing content of 1,4-cyclohexanedimethylene succinate units. This is consistent with the generally accepted idea that the introduction of cyclic moieties into polymer chains alters the macromolecular dynamics by reducing the flexibility. Passing from the oligomers (SPC2 and SPC3) to their chain-extension products (SPC2-C and SPC3-C), an increase in T_g is clearly observed.
- All oligomers are semi-crystalline materials, although they have low melting enthalpies. Their absolute degree of crystallinity is obviously unknown, but in the case of SPC1, which has the lowest content of cyclic units, it approaches

0. For this reason the chain-extended product of this oligomer was not synthesized, since its thermal performance was unsatisfactory.
- SP oligomer shows a bimodal melting endothermic peak, with a maximum melting point of 46.8 °C. SP-C shows an almost identical transition temperature, i.e. 48.0 °C. Furthermore, literature references report a T_m of 115 °C for homopolymeric 1,4-cyclohexanedimethylene succinate. Within the co-oligomeric series, SPC2 exhibits a T_m comparable with that of SP, while SPC3, with the highest content of cyclohexyl units, shows a bimodal melting trace, characterized by a first melting peak centered at 62 °C and a second one centered at 84 °C.

The results obtained show clearly that, although a general improvement in all the thermal properties of the material was obtained, this was relevant only in the presence of high amounts of the cyclic comonomer. This is obviously ascribable to the high intrinsic flexibility of the propylene succinate structure.

3.2.3.3
Segmented Copolymers between Oligo(1,3-Propylene Succinate)s and Poly(Ethylene Glycol)

The thermal and dynamic-mechanical characterization of the chain-extended oligo(propylene succinate)s showed that they are extremely flexible and have thermal properties similar to those of their corresponding oligomers. At the same time, their solubility properties indicate that they are extremely hydrophobic, with a high affinity for non-polar solvents and negligible affinity for polar solvents. This combination of properties makes them eligible for applications as polymeric additives, e.g., as plasticizers, in mass-produced polymeric materials. Oligomeric adipates and succinates have already been proposed for this use. On the other hand, since their degradation products are fully atoxic, they could find useful applications in the biomedical field, e.g., as constituents of drug delivery systems. Therefore, in order to widen the scope and applicability of chain-extended oligo(propylene succinate)s, it was thought convenient to modify their hydrophilic/hydrophobic properties by block copolymerization with hydrophilic PEG (the structure is given in Scheme 13) having molecular weights of 1000 and 2000 *via* the dichloroformate technology [6].

The block copolymerization was carried out following the same general procedure as that used for the chain-extension of α,ω-dihydroxy-terminated oligo(propylene succinate). The molecular weights and molecular weight distributions of the polymers obtained, as well as their transition temperatures, are recorded in Table 7. Solubility tests performed on SP-PEG copolymers obtained

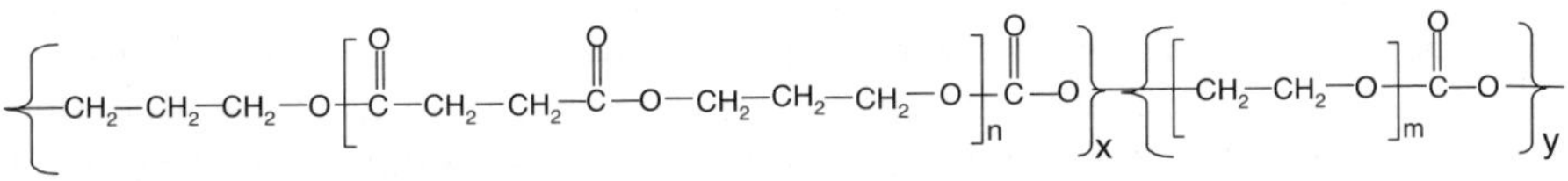

Scheme 13. Segmented copolymer between oligo(propylene succinate) and poly(ethylene glycol)

showed a higher affinity towards polar solvents (in particular low molecular weight alcohols) than oligo(propylene succinate) and its chain-extension products. Moreover, their water absorption capability rises from a few per cent by weight to at least 10 – 15% in the case of copolymers containing 50% PEG.

4
Conclusions

Several synthetic procedures have been developed for the production of biodegradable polyester-based materials from 1,3-propanediol and succinic acid obtainable from renewable resources.

Less useful was the synthetic strategy based on the ROP of lactones deriving from 1,3-propanediol, due to the high cost (low yield) of the starting monomer.

Waxy α,ω-dihydroxy-terminated oligo(propylene succinate)s were successfully transformed into high melting PEU plastomers by a conventional chain extension reaction conducted using MDI as chain-extending agent and different amounts of 1,3-propanediol. These products are semi-crystalline thermoplastic materials, whose crystalline domains stem from the polyurethane segments, and whose T_m varies in the range of 176–208 °C. Their Young's moduli compare well with those of other thermoplastic families used in commodity production. Based on what is generally observed in the case of related materials, we can expect that the segmented PEU copolymers are also biodegradable in different environmental conditions.

A comparison of all the strategies developed indicates that carbonate chain-extended oligomers are the most versatile materials. The chain-extension reaction with phosgene of α, ω-dihydroxy-terminated oligo(propylene succinate)s is an effective way of obtaining high molecular weight poly(ester-carbonate)s. The poly(ester-carbonate)s obtained apparently have a very flexible backbone, resulting in a low glass transition temperature. Their crystalline melting point is also low. As a consequence, their dimensional stability is rapidly lost at temperatures not much higher than room temperature, and this may narrow the field of their practical application.

The design of aliphatic/alicyclic oligomers and chain-extended products was motivated by the need to improve the thermal properties of homopolymeric poly(1,3-propylene succinate)s. The chain-extension reaction of thermally polycondensed oligo(1,3-propylene-co-1,4-cyclohexanedimethylene succinate)s was carried out using phosgene as coupling agent. The results obtained showed that it is feasible to use monomers derived from renewable resources to build up new high molecular weight polymers having aliphatic/alicyclic poly(ester-carbonate) structures. The results obtained demonstrate that the new copolymeric aliphatic/alicyclic poly(ester-carbonate)s also have a very flexible backbone, resulting in low T_g, and that their total crystalline content was also lower than that of the corresponding homopolyester oligomers. Their crystalline melting point increased with increasing alicyclic ester content but their dimensional stability was maintained at temperatures higher than room temperature only in the case of copolymers with a very high content of alicyclic units. We can expect that the poly(ester-carbonate) copolymers are also biodegradable in different environmental conditions.

The polycondensation of oligo(1,3-propylene succinate) with PEG of different lengths using phosgene as coupling agent was proved to be a valuable procedure to obtain high molecular weight hydrophobic/hydrophilic poly(ester-ether-carbonate)s containing biodegradable and biocompatible segments. The poly(ester-ether-carbonate)s obtained are amorphous materials endowed with a very flexible backbone. Their amphiphilic nature is confirmed by their high water swellability. This could make them appealing as constituents of controlled release systems or as compatibilisers.

The results reported demonstrate the feasibility of using monomers derived from renewable resources to build up new polymeric structures endowed with a variety of physical and mechanical properties which make them appealing for practical applications.

References

1. Wilke D (1999) Appl Microbiol Biot 52:135
2. Albertsson AC, Ljungqvist O (1988) Acta polym 39:95
3. Ranucci E, Liu Y, Lindblad MS, Albertsson AC (2000) Macromol Rapid Comm 21:680
4. Liu Y, Lindblad MS, Ranucci E, Albertsson AC (2001) J Polym Sci Pol Chem 39:630
5. Liu Y, Ranucci E, Lindblad MS, Albertsson AC (2001) J Polym Sci Pol Chem 39:2508
6. Liu Y, Ranucci E, Lindblad MS, Albertsson AC (manuscript)
7. Biebl H, Menzel K, Zeng AP, Deckwer WD (1999)Appl Microbiol Biot 52:289
8. Freund A (1881) Monatsh Chem 2:636
9. Deckwer WD (1995) Fems Microbiol Rev 16:143
10. Lee PC, Lee WG, Lee SY, Chang HN (1999) Process Biochem 35:49
11. Nghiem NP, Davison BH, Suttle BE, Richardson GR (1997) Appl Biochem Biotech 63:565
12. Penco M, Donetti R, Mendichi R, Ferruti P (1998) Macromol Chem Physic 199:1737
13. Penco M, Becattini M, Ferruti P, D'Antone S, Deghenghi R (1996) Polym Advan Technol 7:536

Reveiced: April 2001

Author Index Volumes 101–157

Author Index Volumes 1–100 see Volume 100

de, Abajo, J. and *de la Campa, J.G.*: Processable Aromatic Polyimides. Vol. 140, pp. 23-60.
Adolf, D. B. see Ediger, M. D.: Vol. 116, pp. 73-110.
Aharoni, S. M. and *Edwards, S. F.*: Rigid Polymer Networks. Vol. 118, pp. 1-231.
Albertsson, A.-C., Varma, I. K.: Aliphatic Polyesters: Synthesis, Properties and Applications. Vol. 157, pp. 1-40.
Albertsson, A.-C. see Edlund, U.: Vol. 157, pp. 67-112.
Albertsson, A.-C. see Söderqvist Lindblad, M.: Vol. 157, pp. 139-161.
Albertsson, A.-C. see Stridsberg, K. M.: Vol. 157, pp. 41-66.
Améduri, B., Boutevin, B. and *Gramain, P.*: Synthesis of Block Copolymers by Radical Polymerization and Telomerization. Vol. 127, pp. 87-142.
Améduri, B. and *Boutevin, B.*: Synthesis and Properties of Fluorinated Telechelic Monodispersed Compounds. Vol. 102, pp. 133-170.
Amselem, S. see Domb, A. J.: Vol. 107, pp. 93-142.
Andrady, A. L.: Wavelenght Sensitivity in Polymer Photodegradation. Vol. 128, pp. 47-94.
Andreis, M. and *Koenig, J. L.*: Application of Nitrogen-15 NMR to Polymers. Vol. 124, pp. 191-238.
Angiolini, L. see Carlini, C.: Vol. 123, pp. 127-214.
Anseth, K. S., Newman, S. M. and *Bowman, C. N.*: Polymeric Dental Composites: Properties and Reaction Behavior of Multimethacrylate Dental Restorations. Vol. 122, pp. 177-218.
Antonietti, M. see Cölfen, H.: Vol. 150, pp. 67-187.
Armitage, B. A. see O'Brien, D. F.: Vol. 126, pp. 53-58.
Arndt, M. see Kaminski, W.: Vol. 127, pp. 143-187.
Arnold Jr., F. E. and *Arnold, F. E.*: Rigid-Rod Polymers and Molecular Composites. Vol. 117, pp. 257-296.
Arshady, R.: Polymer Synthesis via Activated Esters: A New Dimension of Creativity in Macromolecular Chemistry. Vol. 111, pp. 1-42.

Bahar, I., Erman, B. and *Monnerie, L.*: Effect of Molecular Structure on Local Chain Dynamics: Analytical Approaches and Computational Methods. Vol. 116, pp. 145-206.
Ballauff, M. see Dingenouts, N.: Vol. 144, pp. 1-48.
Baltá-Calleja, F. J., González Arche, A., Ezquerra, T. A., Santa Cruz, C., Batallón, F., Frick, B. and *López Cabarcos, E.*: Structure and Properties of Ferroelectric Copolymers of Poly(vinylidene) Fluoride. Vol. 108, pp. 1-48.
Barnes, M. D. see Otaigbe, J.U.: Vol. 154, pp. 1-86.
Barshtein, G. R. and *Sabsai, O. Y.*: Compositions with Mineralorganic Fillers. Vol. 101, pp.1-28.
Baschnagel, J., Binder, K., Doruker, P., Gusev, A. A., Hahn, O., Kremer, K., Mattice, W. L., Müller-Plathe, F., Murat, M., Paul, W., Santos, S., Sutter, U. W., Tries, V.: Bridging the Gap Between Atomistic and Coarse-Grained Models of Polymers: Status and Perspectives. Vol. 152, pp. 41-156.
Batallán, F. see Baltá-Calleja, F. J.: Vol. 108, pp. 1-48.
Batog, A. E., Pet'ko, I. P., Penczek, P.: Aliphatic-Cycloaliphatic Epoxy Compounds and Polymers. Vol. 144, pp. 49-114.

Barton, J. see Hunkeler, D.: Vol. 112, pp. 115-134.
Bell, C. L. and *Peppas, N. A.*: Biomedical Membranes from Hydrogels and Interpolymer Complexes. Vol. 122, pp. 125-176.
Bellon-Maurel, A. see Calmon-Decriaud, A.: Vol. 135, pp. 207-226.
Bennett, D. E. see O'Brien, D. F.: Vol. 126, pp. 53-84.
Berry, G.C.: Static and Dynamic Light Scattering on Moderately Concentraded Solutions: Isotropic Solutions of Flexible and Rodlike Chains and Nematic Solutions of Rodlike Chains. Vol. 114, pp. 233-290.
Bershtein, V. A. and *Ryzhov, V. A.*: Far Infrared Spectroscopy of Polymers. Vol. 114, pp. 43-122.
Bigg, D. M.: Thermal Conductivity of Heterophase Polymer Compositions. Vol. 119, pp. 1-30.
Binder, K.: Phase Transitions in Polymer Blends and Block Copolymer Melts: Some Recent Developments. Vol. 112, pp. 115-134.
Binder, K.: Phase Transitions of Polymer Blends and Block Copolymer Melts in Thin Films. Vol. 138, pp. 1-90.
Binder, K. see Baschnagel, J.: Vol. 152, pp. 41-156.
Bird, R. B. see Curtiss, C. F.: Vol. 125, pp. 1-102.
Biswas, M. and *Mukherjee, A.*: Synthesis and Evaluation of Metal-Containing Polymers. Vol. 115, pp. 89-124.
Biswas, M. and *Sinha Ray, S.*: Recent Progress in Synthesis and Evaluation of Polymer-Montmorillonite Nanocomposites. Vol. 155, pp. 167-221.
Bolze, J. see Dingenouts, N.: Vol. 144, pp. 1-48.
Boutevin, B. and *Robin, J. J.*: Synthesis and Properties of Fluorinated Diols. Vol. 102. pp. 105-132.
Boutevin, B. see Amédouri, B.: Vol. 102, pp. 133-170.
Boutevin, B. see Améduri, B.: Vol. 127, pp. 87-142.
Bowman, C. N. see Anseth, K. S.: Vol. 122, pp. 177-218.
Boyd, R. H.: Prediction of Polymer Crystal Structures and Properties. Vol. 116, pp. 1-26.
Briber, R. M. see Hedrick, J. L.: Vol. 141, pp. 1-44.
Bronnikov, S. V., Vettegren, V. I. and *Frenkel, S. Y.*: Kinetics of Deformation and Relaxation in Highly Oriented Polymers. Vol. 125, pp. 103-146.
Brown, H. R. see Creton, C.: Vol. 156, pp. 53-135.
Bruza, K. J. see Kirchhoff, R. A.: Vol. 117, pp. 1-66.
Budkowski, A.: Interfacial Phenomena in Thin Polymer Films: Phase Coexistence and Segregation. Vol. 148, pp. 1-112.
Burban, J. H. see Cussler, E. L.: Vol. 110, pp. 67-80.
Burchard, W.: Solution Properties of Branched Macromolecules. Vol. 143, pp. 113-194.

Calmon-Decriaud, A. Bellon-Maurel, V., Silvestre, F.: Standard Methods for Testing the Aerobic Biodegradation of Polymeric Materials. Vol 135, pp. 207-226.
Cameron, N. R. and *Sherrington, D. C.*: High Internal Phase Emulsions (HIPEs)-Structure, Properties and Use in Polymer Preparation. Vol. 126, pp. 163-214.
de la Campa, J. G. see de Abajo, , J.: Vol. 140, pp. 23-60.
Candau, F. see Hunkeler, D.: Vol. 112, pp. 115-134.
Canelas, D. A. and *DeSimone, J. M.*: Polymerizations in Liquid and Supercritical Carbon Dioxide. Vol. 133, pp. 103-140.
Capek, I.: Kinetics of the Free-Radical Emulsion Polymerization of Vinyl Chloride. Vol. 120, pp. 135-206.
Capek, I.: Radical Polymerization of Polyoxyethylene Macromonomers in Disperse Systems. Vol. 145, pp. 1-56.
Capek, I.: Radical Polymerization of Polyoxyethylene Macromonomers in Disperse Systems. Vol. 146, pp. 1-56.
Capek, I. and *Chern, C.-S.*: Radical Polymerization in Direct Mini-Emulsion Systems. Vol. 155, pp. 101-166.
Carlini, C. and *Angiolini, L.*: Polymers as Free Radical Photoinitiators. Vol. 123, pp. 127-214.
Carter, K. R. see Hedrick, J. L.: Vol. 141, pp. 1-44.

Casas-Vazquez, J. see Jou, D.: Vol. 120, pp. 207-266.
Chandrasekhar, V.: Polymer Solid Electrolytes: Synthesis and Structure. Vol 135, pp. 139-206
Chang, J.Y. see Han, M. J.: Vol. 153, pp. 1-36.
Charleux, B., Faust R.: Synthesis of Branched Polymers by Cationic Polymerization. Vol. 142, pp. 1-70.
Chen, P. see Jaffe, M.: Vol. 117, pp. 297-328.
Chern, C.-S. see Capek, I.: Vol. 155, pp. 101-166.
Choe, E.-W. see Jaffe, M.: Vol. 117, pp. 297-328.
Chow, T. S.: Glassy State Relaxation and Deformation in Polymers. Vol. 103, pp. 149-190.
Chung, T.-S. see Jaffe, M.: Vol. 117, pp. 297-328.
Cölfen, H. and *Antonietti, M.*: Field-Flow Fractionation Techniques for Polymer and Colloid Analysis. Vol. 150, pp. 67-187.
Comanita, B. see Roovers, J.: Vol. 142, pp. 179-228.
Connell, J. W. see Hergenrother, P. M.: Vol. 117, pp. 67-110.
Creton, C., Kramer, E. J., Brown, H. R., Hui, C.-Y.: Adhesion and Fracture of Interfaces Between Immiscible Polymers: From the Molecular to the Continuum Scale. Vol. 156, pp. 53-135.
Criado-Sancho, M. see Jou, D.: Vol. 120, pp. 207-266.
Curro, J.G. see Schweizer, K.S.: Vol. 116, pp. 319-378.
Curtiss, C. F. and *Bird, R. B.*: Statistical Mechanics of Transport Phenomena: Polymeric Liquid Mixtures. Vol. 125, pp. 1-102.
Cussler, E. L., Wang, K. L. and *Burban, J. H.*: Hydrogels as Separation Agents. Vol. 110, pp. 67-80.

DeSimone, J. M. see Canelas D. A.: Vol. 133, pp. 103-140.
DiMari, S. see Prokop, A.: Vol. 136, pp. 1-52.
Dimonie, M. V. see Hunkeler, D.: Vol. 112, pp. 115-134.
Dingenouts, N., Bolze, J., Pötschke, D., Ballauf, M.: Analysis of Polymer Latexes by Small-Angle X-Ray Scattering. Vol. 144, pp. 1-48.
Dodd, L. R. and *Theodorou, D. N.*: Atomistic Monte Carlo Simulation and Continuum Mean Field Theory of the Structure and Equation of State Properties of Alkane and Polymer Melts. Vol. 116, pp. 249-282.
Doelker, E.: Cellulose Derivatives. Vol. 107, pp. 199-266.
Dolden, J. G.: Calculation of a Mesogenic Index with Emphasis Upon LC-Polyimides. Vol. 141, pp. 189-245.
Domb, A. J., Amselem, S., Shah, J. and *Maniar, M.*: Polyanhydrides: Synthesis and Characterization. Vol.107, pp. 93-142.
Doruker, P. see Baschnagel, J.: Vol. 152, pp. 41-156.
Dubois, P. see Mecerreyes, D.: Vol. 147, pp. 1-60.
Dubrovskii, S. A. see Kazanskii, K. S.: Vol. 104, pp. 97-134.
Dunkin, I. R. see Steinke, J.: Vol. 123, pp. 81-126.
Dunson, D. L. see McGrath, J. E.: Vol. 140, pp. 61-106.

Eastmond, G. C.: Poly(ε-caprolactone) Blends. Vol.149, pp. 59-223.
Economy, J. and *Goranov, K.*: Thermotropic Liquid Crystalline Polymers for High Performance Applications. Vol. 117, pp. 221-256.
Ediger, M. D. and *Adolf, D. B.*: Brownian Dynamics Simulations of Local Polymer Dynamics. Vol. 116, pp. 73-110.
Edlund, U. Albertsson, A.-C.: Degradable Polymer Microspheres for Controlled Drug Delivery. Vol. 157, pp. 62-112.
Edwards, S. F. see Aharoni, S. M.: Vol. 118, pp. 1-231.
Endo, T. see Yagci, Y.: Vol. 127, pp. 59-86.
Engelhardt, H. and *Grosche, O.*: Capillary Electrophoresis in Polymer Analysis. Vol. 150, pp. 189-217.
Erman, B. see Bahar, I.: Vol. 116, pp. 145-206.
Ewen, B, Richter, D.: Neutron Spin Echo Investigations on the Segmental Dynamics of Polymers in Melts, Networks and Solutions. Vol. 134, pp. 1-130.
Ezquerra, T. A. see Baltá-Calleja, F. J.: Vol. 108, pp. 1-48.

Faust, R. see Charleux, B: Vol. 142, pp. 1-70.
Fekete, E see Pukánszky, B: Vol. 139, pp. 109-154.
Fendler, J.H.: Membrane-Mimetic Approach to Advanced Materials. Vol. 113, pp. 1-209.
Fetters, L. J. see Xu, Z.: Vol. 120, pp. 1-50.
Förster, S. and *Schmidt, M.*: Polyelectrolytes in Solution. Vol. 120, pp. 51-134.
Freire, J. J.: Conformational Properties of Branched Polymers: Theory and Simulations. Vol. 143, pp. 35-112.
Frenkel, S. Y. see Bronnikov, S. V.: Vol. 125, pp. 103-146.
Frick, B. see Baltá-Calleja, F. J.: Vol. 108, pp. 1-48.
Fridman, M. L.: see Terent'eva, J. P.: Vol. 101, pp. 29-64.
Fukui, K. see Otaigbe, J. U.: Vol. 154, pp. 1-86.
Funke, W.: Microgels-Intramolecularly Crosslinked Macromolecules with a Globular Structure. Vol. 136, pp. 137-232.

Galina, H.: Mean-Field Kinetic Modeling of Polymerization: The Smoluchowski Coagulation Equation. Vol. 137, pp. 135-172.
Ganesh, K. see Kishore, K.: Vol. 121, pp. 81-122.
Gaw, K. O. and *Kakimoto, M.*: Polyimide-Epoxy Composites. Vol. 140, pp. 107-136.
Geckeler, K. E. see Rivas, B.: Vol. 102, pp. 171-188.
Geckeler, K. E.: Soluble Polymer Supports for Liquid-Phase Synthesis. Vol. 121, pp. 31-80.
Gehrke, S. H.: Synthesis, Equilibrium Swelling, Kinetics Permeability and Applications of Environmentally Responsive Gels. Vol. 110, pp. 81-144.
de Gennes, P.-G.: Flexible Polymers in Nanopores. Vol. 138, pp. 91-106.
Giannelis, E.P., Krishnamoorti, R., Manias, E.: Polymer-Silicate Nanocomposites: Model Systems for Confined Polymers and Polymer Brushes. Vol. 138, pp. 107-148.
Godovsky, D. Y.: Device Applications of Polymer-Nanocomposites. Vol. 153, pp. 163-205.
Godovsky, D. Y.: Electron Behavior and Magnetic Properties Polymer-Nanocomposites. Vol. 119, pp. 79-122.
González Arche, A. see Baltá-Calleja, F. J.: Vol. 108, pp. 1-48.
Goranov, K. see Economy, J.: Vol. 117, pp. 221-256.
Gramain, P. see Améduri, B.: Vol. 127, pp. 87-142.
Grest, G.S.: Normal and Shear Forces Between Polymer Brushes. Vol. 138, pp. 149-184.
Grigorescu, G, Kulicke, W.-M.: Prediction of Viscoelastic Properties and Shear Stability of Polymers in Solution. Vol. 152, p. 1-40.
Grosberg, A. and *Nechaev, S.*: Polymer Topology. Vol. 106, pp. 1-30.
Grosche, O. see Engelhardt, H.: Vol. 150, pp. 189-217.
Grubbs, R., Risse, W. and *Novac, B.*: The Development of Well-defined Catalysts for Ring-Opening Olefin Metathesis. Vol. 102, pp. 47-72.
van Gunsteren, W. F. see Gusev, A. A.: Vol. 116, pp. 207-248.
Gusev, A. A., Müller-Plathe, F., van Gunsteren, W. F. and *Suter, U. W.*: Dynamics of Small Molecules in Bulk Polymers. Vol. 116, pp. 207-248.
Gusev, A. A. see Baschnagel, J.: Vol. 152, pp. 41-156.
Guillot, J. see Hunkeler, D.: Vol. 112, pp. 115-134.
Guyot, A. and *Tauer, K.*: Reactive Surfactants in Emulsion Polymerization. Vol. 111, pp. 43-66.

Hadjichristidis, N., Pispas, S., Pitsikalis, M., Iatrou, H., Vlahos, C.: Asymmetric Star Polymers Synthesis and Properties. Vol. 142, pp. 71-128.
Hadjichristidis, N. see Xu, Z.: Vol. 120, pp. 1-50.
Hadjichristidis, N. see Pitsikalis, M.: Vol. 135, pp. 1-138.
Hahn, O. see Baschnagel, J.: Vol. 152, pp. 41-156.
Hakkarainen, M.: Aliphatic Polyesters: Abiotic and Biotic Degradation and Degradation Products. Vol. 157, pp. 113-138.
Hall, H. K. see Penelle, J.: Vol. 102, pp. 73-104.
Hamley, I. W.: Crystallization in Block Copolymers. Vol. 148, pp. 113-138.

Hammouda, B.: SANS from Homogeneous Polymer Mixtures: A Unified Overview. Vol. 106, pp. 87-134.
Han, M.J. and Chang, J.Y.: Polynucleotide Analogues. Vol. 153, pp. 1-36.
Harada, A.: Design and Construction of Supramolecular Architectures Consisting of Cyclodextrins and Polymers. Vol. 133, pp. 141-192.
Haralson, M. A. see Prokop, A.: Vol. 136, pp. 1-52.
Hassan, C.M. and Peppas, N.A.: Structure and Applications of Poly(vinyl alcohol) Hydrogels Produced by Conventional Crosslinking or by Freezing/Thawing Methods. Vol. 153, pp. 37-65.
Hawker, C. J. Dentritic and Hyperbranched Macromolecules – Precisely Controlled Macromolecular Architectures. Vol. 147, pp. 113-160.
Hawker, C. J. see Hedrick, J. L.: Vol. 141, pp. 1-44.
Hedrick, J. L., Carter, K. R., Labadie, J. W., Miller, R. D., Volksen, W., Hawker, C. J., Yoon, D. Y., Russell, T. P., McGrath, J. E., Briber, R. M.: Nanoporous Polyimides. Vol. 141, pp. 1-44.
Hedrick, J. L., Labadie, J. W., Volksen, W. and Hilborn, J. G.: Nanoscopically Engineered Polyimides. Vol. 147, pp. 61-112.
Hedrick, J. L. see Hergenrother, P. M.: Vol. 117, pp. 67-110.
Hedrick, J. L. see Kiefer, J.: Vol. 147, pp. 161-247.
Hedrick, J.L. see McGrath, J. E.: Vol. 140, pp. 61-106.
Heller, J.: Poly (Ortho Esters). Vol. 107, pp. 41-92.
Hemielec, A. A. see Hunkeler, D.: Vol. 112, pp. 115-134.
Hergenrother, P. M., Connell, J. W., Labadie, J. W. and Hedrick, J. L.: Poly(arylene ether)s Containing Heterocyclic Units. Vol. 117, pp. 67-110.
Hernández-Barajas, J. see Wandrey, C.: Vol. 145, pp. 123-182.
Hervet, H. see Léger, L.: Vol. 138, pp. 185-226.
Hilborn, J. G. see Hedrick, J. L.: Vol. 147, pp. 61-112.
Hilborn, J. G. see Kiefer, J.: Vol. 147, pp. 161-247.
Hiramatsu, N. see Matsushige, M.: Vol. 125, pp. 147-186.
Hirasa, O. see Suzuki, M.: Vol. 110, pp. 241-262.
Hirotsu, S.: Coexistence of Phases and the Nature of First-Order Transition in Poly-N-isopropylacrylamide Gels. Vol. 110, pp. 1-26.
Höcker, H. see Klee, D.: Vol. 149, pp. 1-57.
Hornsby, P.: Rheology, Compoundind and Processing of Filled Thermoplastics. Vol. 139, pp. 155-216.
Hui, C.-Y. see Creton, C.: Vol. 156, pp. 53-135
Hult, A., Johansson, M., Malmström, E.: Hyperbranched Polymers. Vol. 143, pp. 1-34.
Hunkeler, D., Candau, F., Pichot, C., Hemielec, A. E., Xie, T. Y., Barton, J., Vaskova, V., Guillot, J., Dimonie, M. V., Reichert, K. H.: Heterophase Polymerization: A Physical and Kinetic Comparision and Categorization. Vol. 112, pp. 115-134.
Hunkeler, D. see Prokop, A.: Vol. 136, pp. 1-52; 53-74.
Hunkeler, D see Wandrey, C.: Vol. 145, pp. 123-182.

Iatrou, H. see Hadjichristidis, N.: Vol. 142, pp. 71-128.
Ichikawa, T. see Yoshida, H.: Vol. 105, pp. 3-36.
Ihara, E. see Yasuda, H.: Vol. 133, pp. 53-102.
Ikada, Y. see Uyama, Y.: Vol. 137, pp. 1-40.
Ilavsky, M.: Effect on Phase Transition on Swelling and Mechanical Behavior of Synthetic Hydrogels. Vol. 109, pp. 173-206.
Imai, Y.: Rapid Synthesis of Polyimides from Nylon-Salt Monomers. Vol. 140, pp. 1-23.
Inomata, H. see Saito, S.: Vol. 106, pp. 207-232.
Inoue, S. see Sugimoto, H.: Vol. 146, pp. 39-120.
Irie, M.: Stimuli-Responsive Poly(N-isopropylacrylamide), Photo- and Chemical-Induced Phase Transitions. Vol. 110, pp. 49-66.
Ise, N. see Matsuoka, H.: Vol. 114, pp. 187-232.
Ito, K., Kawaguchi, S,:Poly(macronomers), Homo- and Copolymerization. Vol. 142, pp. 129-178.
Ivanov, A. E. see Zubov, V. P.: Vol. 104, pp. 135-176.

Jacob, S. and Kennedy, J.: Synthesis, Characterization and Properties of OCTA-ARM Polyisobutylene-Based Star Polymers. Vol. 146, pp. 1-38.
Jaffe, M., Chen, P., Choe, E.-W., Chung, T.-S. and *Makhija, S.:* High Performance Polymer Blends. Vol. 117, pp. 297-328.
Jancar, J.: Structure-Property Relationships in Thermoplastic Matrices. Vol. 139, pp. 1-66.
Jerôme, R.: see Mecerreyes, D.: Vol. 147, pp. 1-60.
Jiang, M., Li, M., Xiang, M. and Zhou, H.: Interpolymer Complexation and Miscibility and Enhancement by Hydrogen Bonding. Vol. 146, pp. 121-194.
Jo, W. H. and Yang, J. S.: Molecular Simulation Approaches for Multiphase Polymer Systems. Vol. 156, pp. 1-52.
Johansson, M. see Hult, A.: Vol. 143, pp. 1-34.
Joos-Müller, B. see Funke, W.: Vol. 136, pp. 137-232.
Jou, D., Casas-Vazquez, J. and Criado-Sancho, M.: Thermodynamics of Polymer Solutions under Flow: Phase Separation and Polymer Degradation. Vol. 120, pp. 207-266.

Kaetsu, I.: Radiation Synthesis of Polymeric Materials for Biomedical and Biochemical Applications. Vol. 105, pp. 81-98.
Kaji, K. see Kanaya, T.: Vol. 154, pp. 87-141.
Kakimoto, M. see Gaw, K. O.: Vol. 140, pp. 107-136.
Kaminski, W. and Arndt, M.: Metallocenes for Polymer Catalysis. Vol. 127, pp. 143-187.
Kammer, H. W., Kressler, H. and Kummerloewe, C.: Phase Behavior of Polymer Blends - Effects of Thermodynamics and Rheology. Vol. 106, pp. 31-86.
Kanaya, T. and Kaji, K.: Dynamcis in the Glassy State and Near the Glass Transition of Amorphous Polymers as Studied by Neutron Scattering. Vol. 154, pp. 87-141.
Kandyrin, L. B. and Kuleznev, V. N.: The Dependence of Viscosity on the Composition of Concentrated Dispersions and the Free Volume Concept of Disperse Systems. Vol. 103, pp. 103-148.
Kaneko, M. see Ramaraj, R.: Vol. 123, pp. 215-242.
Kang, E. T., Neoh, K. G. and Tan, K. L.: X-Ray Photoelectron Spectroscopic Studies of Electroactive Polymers. Vol. 106, pp. 135-190.
Karlsson, S. see Söderqvist Lindblad, M.: Vol. 157, pp. 139–161.
Kato, K. see Uyama, Y.: Vol. 137, pp. 1-40.
Kawaguchi, S. see Ito, K.: Vol. 142, p 129-178.
Kazanskii, K. S. and Dubrovskii, S. A.: Chemistry and Physics of „Agricultural" Hydrogels. Vol. 104, pp. 97-134.
Kennedy, J. P. see Jacob, S.: Vol. 146, pp. 1-38.
Kennedy, J. P. see Majoros, I.: Vol. 112, pp. 1-113.
Khokhlov, A., Starodybtzev, S. and Vasilevskaya, V.: Conformational Transitions of Polymer Gels: Theory and Experiment. Vol. 109, pp. 121-172.
Kiefer, J., Hedrick J. L. and Hiborn, J. G.: Macroporous Thermosets by Chemically Induced Phase Separation. Vol. 147, pp. 161-247.
Kilian, H. G. and Pieper, T.: Packing of Chain Segments. A Method for Describing X-Ray Patterns of Crystalline, Liquid Crystalline and Non-Crystalline Polymers. Vol. 108, pp. 49-90.
Kim, J. see Quirk, R.P.: Vol. 153, pp. 67-162.
Kishore, K. and Ganesh, K.: Polymers Containing Disulfide, Tetrasulfide, Diselenide and Ditelluride Linkages in the Main Chain. Vol. 121, pp. 81-122.
Kitamaru, R.: Phase Structure of Polyethylene and Other Crystalline Polymers by Solid-State ^{13}C/MNR. Vol. 137, pp 41-102.
Klee, D. and Höcker, H.: Polymers for Biomedical Applications: Improvement of the Interface Compatibility. Vol. 149, pp. 1-57.
Klier, J. see Scranton, A. B.: Vol. 122, pp. 1-54.
Kobayashi, S., Shoda, S. and Uyama, H.: Enzymatic Polymerization and Oligomerization. Vol. 121, pp. 1-30.
Köhler, W. and Schäfer, R.: Polymer Analysis by Thermal-Diffusion Forced Rayleigh Scattering. Vol. 151, pp. 1-59.

Koenig, J. L. see Andreis, M.: Vol. 124, pp. 191-238.
Koike, T.: Viscoelastic Behavior of Epoxy Resins Before Crosslinking. Vol. 148, pp. 139-188.
Kokufuta, E.: Novel Applications for Stimulus-Sensitive Polymer Gels in the Preparation of Functional Immobilized Biocatalysts. Vol. 110, pp. 157-178.
Konno, M. see Saito, S.: Vol. 109, pp. 207-232.
Kopecek, J. see Putnam, D.: Vol. 122, pp. 55-124.
Koßmehl, G. see Schopf, G.: Vol. 129, pp. 1-145.
Kramer, E. J. see Creton, C.: Vol. 156, pp. 53-135.
Kremer, K. see Baschnagel, J.: Vol. 152, pp. 41-156.
Kressler, J. see Kammer, H. W.: Vol. 106, pp. 31-86.
Kricheldorf, H. R.: Liquid-Cristalline Polyimides. Vol. 141, pp. 83-188.
Krishnamoorti, R. see Giannelis, E.P.: Vol. 138, pp. 107-148.
Kirchhoff, R. A. and *Bruza, K. J.*: Polymers from Benzocyclobutenes. Vol. 117, pp. 1-66.
Kuchanov, S. I.: Modern Aspects of Quantitative Theory of Free-Radical Copolymerization. Vol. 103, pp. 1-102.
Kuchanov, S. I.: Principles of Quantitive Description of Chemical Structure of Synthetic Polymers. Vol. 152, p. 157-202.
Kudaibergennow, S.E.: Recent Advances in Studying of Synthetic Polyampholytes in Solutions. Vol. 144, pp. 115-198.
Kuleznev, V. N. see Kandyrin, L. B.: Vol. 103, pp. 103-148.
Kulichkhin, S. G. see Malkin, A. Y.: Vol. 101, pp. 217-258.
Kulicke, W.-M. see Grigorescu, G.: Vol. 152, p. 1-40.
Kummerloewe, C. see Kammer, H. W.: Vol. 106, pp. 31-86.
Kuznetsova, N. P. see Samsonov, G. V.: Vol. 104, pp. 1-50. Labadie, J. W. see Hergenrother, P. M.: Vol. 117, pp. 67-110.

Labadie, J. W. see Hedrick, J. L.: Vol. 141, pp. 1-44.
Labadie, J. W. see Hedrick, J. L.: Vol. 147, pp. 61-112.
Lamparski, H. G. see O´Brien, D. F.: Vol. 126, pp. 53-84.
Laschewsky, A.: Molecular Concepts, Self-Organisation and Properties of Polysoaps. Vol. 124, pp. 1-86.
Laso, M. see Leontidis, E.: Vol. 116, pp. 283-318.
Lazár, M. and *RychlΩ, R.*: Oxidation of Hydrocarbon Polymers. Vol. 102, pp. 189-222.
Lechowicz, J. see Galina, H.: Vol. 137, pp. 135-172.
Léger, L., Raphaël, E., Hervet, H.: Surface-Anchored Polymer Chains: Their Role in Adhesion and Friction. Vol. 138, pp. 185-226.
Lenz, R. W.: Biodegradable Polymers. Vol. 107, pp. 1-40.
Leontidis, E., de Pablo, J. J., Laso, M. and *Suter, U. W.*: A Critical Evaluation of Novel Algorithms for the Off-Lattice Monte Carlo Simulation of Condensed Polymer Phases. Vol. 116, pp. 283-318.
Lee, B. see Quirk, R.P: Vol. 153, pp. 67-162.
Lee, Y. see Quirk, R.P: Vol. 153, pp. 67-162.
Lesec, J. see Viovy, J.-L.: Vol. 114, pp. 1-42.
Li, M. see Jiang, M.: Vol. 146, pp. 121-194.
Liang, G. L. see Sumpter, B. G.: Vol. 116, pp. 27-72.
Lienert, K.-W.: Poly(ester-imide)s for Industrial Use. Vol. 141, pp. 45-82.
Lin, J. and *Sherrington, D. C.*: Recent Developments in the Synthesis, Thermostability and Liquid Crystal Properties of Aromatic Polyamides. Vol. 111, pp. 177-220.
Liu, Y. see Söderqvist Lindblad, M.: Vol. 157, pp. 139–161
López Cabarcos, E. see Baltá-Calleja, F. J.: Vol. 108, pp. 1-48.

Majoros, I., Nagy, A. and *Kennedy, J. P.*: Conventional and Living Carbocationic Polymerizations United. I. A Comprehensive Model and New Diagnostic Method to Probe the Mechanism of Homopolymerizations. Vol. 112, pp. 1-113.
Makhija, S. see Jaffe, M.: Vol. 117, pp. 297-328.
Malmström, E. see Hult, A.: Vol. 143, pp. 1-34.

Malkin, A. Y. and *Kulichkhin, S. G.*: Rheokinetics of Curing. Vol. 101, pp. 217-258.
Maniar, M. see Domb, A. J.: Vol. 107, pp. 93-142.
Manias, E., see Giannelis, E.P.: Vol. 138, pp. 107-148.
Mashima, K., Nakayama, Y. and *Nakamura, A.*: Recent Trends in Polymerization of a-Olefins Catalyzed by Organometallic Complexes of Early Transition Metals. Vol. 133, pp. 1-52.
Mathew, D. see Reghunadhan Nair, C. P.: Vol. 155, pp. 1-99.
Matsumoto, A.: Free-Radical Crosslinking Polymerization and Copolymerization of Multivinyl Compounds. Vol. 123, pp. 41-80.
Matsumoto, A. see Otsu, T.: Vol. 136, pp. 75-138.
Matsuoka, H. and *Ise, N.*: Small-Angle and Ultra-Small Angle Scattering Study of the Ordered Structure in Polyelectrolyte Solutions and Colloidal Dispersions. Vol. 114, pp. 187-232.
Matsushige, K., Hiramatsu, N. and *Okabe, H.*: Ultrasonic Spectroscopy for Polymeric Materials. Vol. 125, pp. 147-186.
Mattice, W. L. see Rehahn, M.: Vol. 131/132, pp. 1-475.
Mattice, W. L. see Baschnagel, J.: Vol. 152, p. 41-156.
Mays, W. see Xu, Z.: Vol. 120, pp. 1-50.
Mays, J.W. see Pitsikalis, M.: Vol.135, pp. 1-138.
McGrath, J. E. see Hedrick, J. L.: Vol. 141, pp. 1-44.
McGrath, J. E., Dunson, D. L., Hedrick, J. L.: Synthesis and Characterization of Segmented Polyimide-Polyorganosiloxane Copolymers. Vol. 140, pp. 61-106.
McLeish, T.C. B., Milner, S. T.: Entangled Dynamics and Melt Flow of Branched Polymers. Vol. 143, pp. 195-256.
Mecerreyes, D., Dubois, P. and *Jerôme, R.*: Novel Macromolecular Architectures Based on Aliphatic Polyesters: Relevance of the „Coordination-Insertion" Ring-Opening Polymerization. Vol. 147, pp. 1 -60.
Mecham, S. J. see McGrath, J. E.: Vol. 140, pp. 61-106.
Mikos, A. G. see Thomson, R. C.: Vol. 122, pp. 245-274.
Milner, S. T. see McLeish, T. C. B.: Vol. 143, pp. 195-256.
Mison, P. and *Sillion, B.*: Thermosetting Oligomers Containing Maleimides and Nadiimides End-Groups. Vol. 140, pp. 137-180.
Miyasaka, K.: PVA-Iodine Complexes: Formation, Structure and Properties. Vol. 108. pp. 91-130.
Miller, R. D. see Hedrick, J. L.: Vol. 141, pp. 1-44.
Monnerie, L. see Bahar, I.: Vol. 116, pp. 145-206.
Morishima, Y.: Photoinduced Electron Transfer in Amphiphilic Polyelectrolyte Systems. Vol. 104, pp. 51-96.
Morton M. see Quirk, R.P: Vol. 153, pp. 67-162
Mours, M. see Winter, H. H.: Vol. 134, pp. 165-234.
Müllen, K. see Scherf, U.: Vol. 123, pp. 1-40.
Müller-Plathe, F. see Gusev, A. A.: Vol. 116, pp. 207-248.
Müller-Plathe, F. see Baschnagel, J.: Vol. 152, p. 41-156.
Mukerherjee, A. see Biswas, M.: Vol. 115, pp. 89-124.
Murat, M. see Baschnagel, J.: Vol. 152, p. 41-156.
Mylnikov, V.: Photoconducting Polymers. Vol. 115, pp. 1-88.

Nagy, A. see Majoros, I.: Vol. 112, pp. 1-11.
Nakamura, A. see Mashima, K.: Vol. 133, pp. 1-52.
Nakayama, Y. see Mashima, K.: Vol. 133, pp. 1-52.
Narasinham, B., Peppas, N. A.: The Physics of Polymer Dissolution: Modeling Approaches and Experimental Behavior. Vol. 128, pp. 157-208.
Nechaev, S. see Grosberg, A.: Vol. 106, pp. 1-30.
Neoh, K. G. see Kang, E. T.: Vol. 106, pp. 135-190.
Newman, S. M. see Anseth, K. S.: Vol. 122, pp. 177-218.
Nijenhuis, K. te: Thermoreversible Networks. Vol. 130, pp. 1-252.
Ninan, K.N. see Reghunadhan Nair, C. P.: Vol. 155, pp. 1-99.
Noid, D. W. see Otaigbe, J.U.: Vol. 154, pp. 1-86.
Noid, D. W. see Sumpter, B. G.: Vol. 116, pp. 27-72.

Novac, B. see Grubbs, R.: Vol. 102, pp. 47-72.
Novikov, V. V. see Privalko, V. P.: Vol. 119, pp. 31-78.

O'Brien, D. F., Armitage, B. A., Bennett, D. E. and *Lamparski, H. G.*: Polymerization and Domain Formation in Lipid Assemblies. Vol. 126, pp. 53-84.
Ogasawara, M.: Application of Pulse Radiolysis to the Study of Polymers and Polymerizations. Vol.105, pp. 37-80.
Okabe, H. see Matsushige, K.: Vol. 125, pp. 147-186.
Okada, M.: Ring-Opening Polymerization of Bicyclic and Spiro Compounds. Reactivities and Polymerization Mechanisms. Vol. 102, pp. 1-46.
Okano, T.: Molecular Design of Temperature-Responsive Polymers as Intelligent Materials. Vol. 110, pp. 179-198.
Okay, O. see Funke, W.: Vol. 136, pp. 137-232.
Onuki, A.: Theory of Phase Transition in Polymer Gels. Vol. 109, pp. 63-120.
Osad'ko, I.S.: Selective Spectroscopy of Chromophore Doped Polymers and Glasses. Vol. 114, pp. 123-186.
Otaigbe, J. U., Barnes, M. D., Fukui, K., Sumpter, B. G., Noid, D. W.: Generation, Characterization, and Modeling of Polymer Micro- and Nano-Particles. Vol. 154, pp. 1-86.
Otsu, T., Matsumoto, A.: Controlled Synthesis of Polymers Using the Iniferter Technique: Developments in Living Radical Polymerization. Vol. 136, pp. 75-138.

de Pablo, J. J. see Leontidis, E.: Vol. 116, pp. 283-318.
Padias, A. B. see Penelle, J.: Vol. 102, pp. 73-104.
Pascault, J.-P. see Williams, R. J. J.: Vol. 128, pp. 95-156.
Pasch, H.: Analysis of Complex Polymers by Interaction Chromatography. Vol. 128, pp. 1-46.
Pasch, H.: Hyphenated Techniques in Liquid Chromatography of Polymers. Vol. 150, pp. 1-66.
Paul, W. see Baschnagel, J.: Vol. 152, p. 41-156.
Penczek, P. see Batog, A. E.: Vol. 144, pp. 49-114.
Penelle, J., Hall, H. K., Padias, A. B. and *Tanaka, H.*: Captodative Olefins in Polymer Chemistry. Vol. 102, pp. 73-104.
Peppas, N. A. see Bell, C. L.: Vol. 122, pp. 125-176.
Peppas, N.A. see Hassan, C.M.: Vol. 153, pp. 37-65
Peppas, N. A. see Narasimhan, B.: Vol. 128, pp. 157-208.
Pet'ko, I. P. see Batog, A. E.: Vol. 144, pp. 49-114.
Pichot, C. see Hunkeler, D.: Vol. 112, pp. 115-134.
Pieper, T. see Kilian, H. G.: Vol. 108, pp. 49-90.
Pispas, S. see Pitsikalis, M.: Vol. 135, pp. 1-138.
Pispas, S. see Hadjichristidis: Vol. 142, pp. 71-128.
Pitsikalis, M., Pispas, S., Mays, J. W., Hadjichristidis, N.: Nonlinear Block Copolymer Architectures. Vol. 135, pp. 1-138.
Pitsikalis, M. see Hadjichristidis: Vol. 142, pp. 71-128.
Pötschke, D. see Dingenouts, N.: Vol 144, pp. 1-48.
Pokrovskii, V. N.: The Mesoscopic Theory of the Slow Relaxation of Linear Macromolecules. Vol. 154, pp. 143-219.
Pospíšil, J.: Functionalized Oligomers and Polymers as Stabilizers for Conventional Polymers. Vol. 101, pp. 65-168.
Pospíšil, J.: Aromatic and Heterocyclic Amines in Polymer Stabilization. Vol. 124, pp. 87-190.
Powers, A. C. see Prokop, A.: Vol. 136, pp. 53-74.
Priddy, D. B.: Recent Advances in Styrene Polymerization. Vol. 111, pp. 67-114.
Priddy, D. B.: Thermal Discoloration Chemistry of Styrene-co-Acrylonitrile. Vol. 121, pp. 123-154.
Privalko, V. P. and *Novikov, V. V.*: Model Treatments of the Heat Conductivity of Heterogeneous Polymers. Vol. 119, pp 31-78.
Prokop, A., Hunkeler, D., Powers, A. C., Whitesell, R. R., Wang, T. G.: Water Soluble Polymers for Immunoisolation II: Evaluation of Multicomponent Microencapsulation Systems. Vol. 136, pp. 53-74.

Prokop, A., Hunkeler, D., DiMari, S., Haralson, M. A., Wang, T. G.: Water Soluble Polymers for Immunoisolation I: Complex Coacervation and Cytotoxicity. Vol. 136, pp. 1-52.
Pukánszky, B. and Fekete, E.: Adhesion and Surface Modification. Vol. 139, pp. 109-154.
Putnam, D. and *Kopecek, J.*: Polymer Conjugates with Anticancer Acitivity. Vol. 122, pp. 55- 124.

Quirk, R.P. and Yoo, T., Lee, Y., M., Kim, J. and Lee, B.: Applications of 1,1-Diphenylethylene Chemistry in Anionic Synthesis of Polymers with Controlled Structures. Vol. 153, pp. 67-162.

Ramaraj, R. and *Kaneko, M.*: Metal Complex in Polymer Membrane as a Model for Photosynthetic Oxygen Evolving Center. Vol. 123, pp. 215-242.
Rangarajan, B. see Scranton, A. B.: Vol. 122, pp. 1-54.
Ranucci, E. see Söderqvist Lindblad, M.: Vol. 157, pp. 139–161.
Raphaël, E. see Léger, L.: Vol. 138, pp. 185-226.
Reddinger, J. L. and *Reynolds, J. R.*: Molecular Engineering of π-Conjugated Polymers. Vol. 145, pp. 57-122.
Reghunadhan Nair, C.P., Mathew, D. and *Ninan, K.N.,* : Cyanate Ester Resins, Recent Developments. Vol. 155, pp. 1-99.
Reichert, K. H. see Hunkeler, D.: Vol. 112, pp. 115-134.
Rehahn, M., Mattice, W. L., Suter, U. W.: Rotational Isomeric State Models in Macromolecular Systems. Vol. 131/132, pp. 1-475.
Reynolds, J.R. see Reddinger, J. L.: Vol. 145, pp. 57-122.
Richter, D. see Ewen, B.: Vol. 134, pp.1-130.
Risse, W. see Grubbs, R.: Vol. 102, pp. 47-72.
Rivas, B. L. and *Geckeler, K. E.*: Synthesis and Metal Complexation of Poly(ethyleneimine) and Derivatives. Vol. 102, pp. 171-188.
Robin, J. J. see Boutevin, B.: Vol. 102, pp. 105-132.
Roe, R.-J.: MD Simulation Study of Glass Transition and Short Time Dynamics in Polymer Liquids. Vol. 116, pp. 111-114.
Roovers, J., Comanita, B.: Dendrimers and Dendrimer-Polymer Hybrids. Vol. 142, pp 179-228.
Rothon, R. N.: Mineral Fillers in Thermoplastics: Filler Manufacture and Characterisation. Vol. 139, pp. 67-108.
Rozenberg, B. A. see Williams, R. J. J.: Vol. 128, pp. 95-156.
Ruckenstein, E.: Concentrated Emulsion Polymerization. Vol. 127, pp. 1-58.
Rusanov, A. L.: Novel Bis (Naphtalic Anhydrides) and Their Polyheteroarylenes with Improved Processability. Vol. 111, pp. 115-176.
Russel, T. P. see Hedrick, J. L.: Vol. 141, pp. 1-44.
Rychlý, J. see Lazár, M.: Vol. 102, pp. 189-222.
Ryner, M. see Stridsberg, K. M.: Vol. 157, pp. 41-66.
Ryzhov, V. A. see Bershtein, V. A.: Vol. 114, pp. 43-122.

Sabsai, O. Y. see Barshtein, G. R.: Vol. 101, pp. 1-28.
Saburov, V. V. see Zubov, V. P.: Vol. 104, pp. 135-176.
Saito, S., Konno, M. and *Inomata, H.*: Volume Phase Transition of N-Alkylacrylamide Gels. Vol. 109, pp. 207-232.
Samsonov, G. V. and *Kuznetsova, N. P.*: Crosslinked Polyelectrolytes in Biology. Vol. 104, pp. 1-50.
Santa Cruz, C. see Baltá-Calleja, F. J.: Vol. 108, pp. 1-48.
Santos, S. see Baschnagel, J.: Vol. 152, p. 41-156.
Sato, T. and *Teramoto, A.*: Concentrated Solutions of Liquid-Christalline Polymers. Vol. 126, pp. 85-162.
Schäfer R. see Köhler, W.: Vol. 151, pp. 1-59.
Scherf, U. and *Müllen, K.*: The Synthesis of Ladder Polymers. Vol. 123, pp. 1-40.
Schmidt, M. see Förster, S.: Vol. 120, pp. 51-134.

Schopf, G. and *Koßmehl, G.:* Polythiophenes - Electrically Conductive Polymers. Vol. 129, pp. 1-145.
Schweizer, K. S.: Prism Theory of the Structure, Thermodynamics, and Phase Transitions of Polymer Liquids and Alloys. Vol. 116, pp. 319-378.
Scranton, A. B., Rangarajan, B. and *Klier, J.:* Biomedical Applications of Polyelectrolytes. Vol. 122, pp. 1-54.
Sefton, M. V. and *Stevenson, W. T. K.:* Microencapsulation of Live Animal Cells Using Polycrylates. Vol.107, pp. 143-198.
Shamanin, V. V.: Bases of the Axiomatic Theory of Addition Polymerization. Vol. 112, pp. 135-180.
Sheiko, S. S.: Imaging of Polymers Using Scanning Force Microscopy: From Superstructures to Individual Molecules. Vol. 151, pp. 61-174.
Sherrington, D. C. see Cameron, N. R., Vol. 126, pp. 163-214.
Sherrington, D. C. see Lin, J.: Vol. 111, pp. 177-220.
Sherrington, D. C. see Steinke, J.: Vol. 123, pp. 81-126.
Shibayama, M. see Tanaka, T.: Vol. 109, pp. 1-62.
Shiga, T.: Deformation and Viscoelastic Behavior of Polymer Gels in Electric Fields. Vol. 134, pp. 131-164.
Shoda, S. see Kobayashi, S.: Vol. 121, pp. 1-30.
Siegel, R. A.: Hydrophobic Weak Polyelectrolyte Gels: Studies of Swelling Equilibria and Kinetics. Vol. 109, pp. 233-268.
Silvestre, F. see Calmon-Decriaud, A.: Vol. 207, pp. 207-226.
Sillion, B. see Mison, P.: Vol. 140, pp. 137-180.
Singh, R. P. see Sivaram, S.: Vol. 101, pp. 169-216.
Sinha Ray, S. see Biswas, M: Vol. 155, pp. 167-221.
Sivaram, S. and *Singh, R. P:* Degradation and Stabilization of Ethylene-Propylene Copolymers and Their Blends: A Critical Review. Vol. 101, pp. 169-216.
Söderqvist Lindblad, M., Liu, Y., Albertsson, A.-C., Ranucci, E., Karlsson, S.: Polymer from Renewable Resources. Vol. 157, pp. 139-161
Starodybtzev, S. see Khokhlov, A.: Vol. 109, pp. 121-172.
Steinke, J., Sherrington, D. C. and *Dunkin, I. R.:* Imprinting of Synthetic Polymers Using Molecular Templates. Vol. 123, pp. 81-126.
Stenzenberger, H. D.: Addition Polyimides. Vol. 117, pp. 165-220.
Stevenson, W. T. K. see Sefton, M. V.: Vol. 107, pp. 143-198.
Stridsberg, K. M., Ryner, M., Albertsson, A.-C.: Controlled Ring-Opening Polymerization: Polymers with Designed Macromolecular Architecture. Vol. 157, pp. 41-66.
Suematsu, K.: Recent Progress of Gel Theory: Ring, Excluded Volume, and Dimension. Vol. 156, pp. 136-214.
Sumpter, B. G., Noid, D. W., Liang, G. L. and *Wunderlich, B.:* Atomistic Dynamics of Macromolecular Crystals. Vol. 116, pp. 27-72.
Sumpter, B. G. see Otaigbe, J.U.: Vol. 154, pp. 1-86.
Sugimoto, H. and *Inoue, S.:* Polymerization by Metalloporphyrin and Related Complexes. Vol. 146, pp. 39-120.
Suter, U. W. see Gusev, A. A.: Vol. 116, pp. 207-248.
Suter, U. W. see Leontidis, E.: Vol. 116, pp. 283-318.
Suter, U. W. see Rehahn, M.: Vol. 131/132, pp. 1-475.
Suter, U. W. see Baschnagel, J.: Vol. 152, p. 41-156.
Suzuki, A.: Phase Transition in Gels of Sub-Millimeter Size Induced by Interaction with Stimuli. Vol. 110, pp. 199-240.
Suzuki, A. and *Hirasa, O.:* An Approach to Artifical Muscle by Polymer Gels due to Micro-Phase Separation. Vol. 110, pp. 241-262.

Tagawa, S.: Radiation Effects on Ion Beams on Polymers. Vol. 105, pp. 99-116.
Tan, K. L. see Kang, E. T.: Vol. 106, pp. 135-190.

Tanaka, H. and *Shibayama, M.*: Phase Transition and Related Phenomena of Polymer Gels. Vol. 109, pp. 1-62.
Tanaka, T. see Penelle, J.: Vol. 102, pp. 73-104.
Tauer, K. see Guyot, A.: Vol. 111, pp. 43-66.
Teramoto, A. see Sato, T.: Vol. 126, pp. 85-162.
Terent'eva, J. P. and *Fridman, M. L.*: Compositions Based on Aminoresins. Vol. 101, pp. 29-64.
Theodorou, D. N. see Dodd, L. R.: Vol. 116, pp. 249-282.
Thomson, R. C., Wake, M. C., Yaszemski, M. J. and *Mikos, A. G.*: Biodegradable Polymer Scaffolds to Regenerate Organs. Vol. 122, pp. 245-274.
Tokita, M.: Friction Between Polymer Networks of Gels and Solvent. Vol. 110, pp. 27-48.
Tries, V. see Baschnagel, J:. Vol. 152, p. 41-156.
Tsuruta, T.: Contemporary Topics in Polymeric Materials for Biomedical Applications. Vol. 126, pp. 1-52.

Uyama, H. see Kobayashi, S.: Vol. 121, pp. 1-30.
Uyama, Y: Surface Modification of Polymers by Grafting. Vol. 137, pp. 1-40.

Varma, I. K. see Albertsson, A.-C.: Vol. 157, pp. 1-40.
Vasilevskaya, V. see Khokhlov, A.: Vol. 109, pp. 121-172.
Vaskova, V. see Hunkeler, D.: Vol.:112, pp. 115-134.
Verdugo, P.: Polymer Gel Phase Transition in Condensation-Decondensation of Secretory Products. Vol. 110, pp. 145-156.
Vettegren, V. I.: see Bronnikov, S. V.: Vol. 125, pp. 103-146.
Viovy, J.-L. and *Lesec, J.*: Separation of Macromolecules in Gels: Permeation Chromatography and Electrophoresis. Vol. 114, pp. 1-42.
Vlahos, C. see Hadjichristidis, N.: Vol. 142, pp. 71-128.
Volksen, W.: Condensation Polyimides: Synthesis, Solution Behavior, and Imidization Characteristics. Vol. 117, pp. 111-164.
Volksen, W. see Hedrick, J. L.: Vol. 141, pp. 1-44.
Volksen, W. see Hedrick, J. L.: Vol. 147, pp. 61-112.

Wake, M. C. see Thomson, R. C.: Vol. 122, pp. 245-274.
Wandrey C., Hernández-Barajas, J. and *Hunkeler, D.*: Diallyldimethylammonium Chloride and its Polymers. Vol. 145, pp. 123-182.
Wang, K. L. see Cussler, E. L.: Vol. 110, pp. 67-80.
Wang, S.-Q.: Molecular Transitions and Dynamics at Polymer/Wall Interfaces: Origins of Flow Instabilities and Wall Slip. Vol. 138, pp. 227-276.
Wang, T. G. see Prokop, A.: Vol. 136, pp.1-52; 53-74.
Whitesell, R. R. see Prokop, A.: Vol. 136, pp. 53-74.
Williams, R. J. J., Rozenberg, B. A., Pascault, J.-P.: Reaction Induced Phase Separation in Modified Thermosetting Polymers. Vol. 128, pp. 95-156.
Winter, H. H., Mours, M.: Rheology of Polymers Near Liquid-Solid Transitions. Vol. 134, pp. 165-234.
Wu, C.: Laser Light Scattering Characterization of Special Intractable Macromolecules in Solution. Vol 137, pp. 103-134.
Wunderlich, B. see Sumpter, B. G.: Vol. 116, pp. 27-72.

Xiang, M. see Jiang, M.: Vol. 146, pp. 121-194.
Xie, T. Y. see Hunkeler, D.: Vol. 112, pp. 115-134.
Xu, Z., Hadjichristidis, N., Fetters, L. J. and *Mays, J. W.*: Structure/Chain-Flexibility Relationships of Polymers. Vol. 120, pp. 1-50.

Yagci, Y. and *Endo, T.*: N-Benzyl and N-Alkoxy Pyridium Salts as Thermal and Photochemical Initiators for Cationic Polymerization. Vol. 127, pp. 59-86.
Yannas, I. V.: Tissue Regeneration Templates Based on Collagen-Glycosaminoglycan Copolymers. Vol. 122, pp. 219-244.
Yang, J. S. see Jo, W. H.: Vol. 156, pp. 1-52.
Yamaoka, H.: Polymer Materials for Fusion Reactors. Vol. 105, pp. 117-144.

Yasuda, H. and *Ihara, E.*: Rare Earth Metal-Initiated Living Polymerizations of Polar and Nonpolar Monomers. Vol. 133, pp. 53-102.
Yaszemski, M. J. see Thomson, R. C.: Vol. 122, pp. 245-274.
Yoo, T. see Quirk, R.P.: Vol. 153, pp. 67-162.
Yoon, D. Y. see Hedrick, J. L.: Vol. 141, pp. 1-44.
Yoshida, H. and *Ichikawa, T.*: Electron Spin Studies of Free Radicals in Irradiated Polymers. Vol. 105, pp. 3-36.

Zhou, H. see Jiang, M.: Vol. 146, pp. 121-194.
Zubov, V. P., Ivanov, A. E. and *Saburov, V. V.*: Polymer-Coated Adsorbents for the Separation of Biopolymers and Particles. Vol. 104, pp. 135-176.

Subject Index

Acetic acid 127, 130
Administration 69, 72–75, 83, 101, 105
Aggregation 50, 53
Aliphatic polyesters 42
Alkaline hydrolysis 131
Aluminum isopropoxide 49
Amino acid biosynthesis 143–144
Anaerobic sludge 123
Anhydride 86, 88–95, 104
Anionic polymerization 11, 45
Autocatalysis 78, 92, 94, 95
Autoclaving 102

Back-biting 46
Bioassimilation 128, 130
Biocompatibility 3, 70, 76, 77, 80–88, 96, 99, 105
Biocompatible material 60
Biodegradability 3
Biodegradable polymers 58
Biodegradation, poly(3-hydroxybutyrate) 124
–, polycaprolactone 121–123
–, polylactide 119–121
Bionolle 3, 5, 26, 31
Bioresorption 3, 76, 81
Biostability 77
Biosynthesis 117
Bis(β-lactones) 12
Blend 78, 81, 84, 86, 92, 93, 97, 98, 104
Blending 115, 117
Block copolymers 48, 55
Bulk disintegration 118
Burst 100
2-Butenoic acid 131
Butyric acid 130

Capillary zone electrophoresis 126, 128, 130
Caprolactone 129
– Caprolactone 78–81, 84, 86, 92, 97

Carbocationic polymerization 15
Catalysts 149, 151
Cationic polymerization 44
Cellulose 68, 77
Chain extension 148–160
Chitin 77
Chitosan 77
Coacervation 98
Composting 121, 123
Coordination ring-opening polymerization of lactone 16
Coordination-Insertion mechanism 45
Copolymer 78, 81–88, 92, 96–98
Copolymerization 115, 117
Crosslinking 123
Cyclic dimer 149–150
Cyclic ketene acetals 12
Cyclic tin alkoxides 50
Cyclohexanedimethanol 148, 151, 157–159

Degradation 70–105, 141, 147–148, 159
–, in vitro 124
–, in natural environments 121–124
Degradation products 125–131
   extraction 125, 126
Degradation rate 119
Dendrimer 24
Dexon 3
Dichloroformate synthesis 151, 155–159
Diffusion 69–75, 87, 100
Diisocyanate synthesis 152
Diisophenylmethane diisocyanate (MDI) 151–152
Dioxepan-one 149
Disintegration, bulk 118
Drug delivery systems 141, 159
Dynamic mechanical measurements 153–156, 159

Enzymatic degradation 70–92, 119, 121–124
Enzymatic essay 126, 128

Erosion 69, 71–75, 78, 86–88, 90, 92–95, 103, 105
Ethyl ester of lactoyl lactic acid 127
Ethylene oxide 102, 103

FDA 70, 77, 80–82, 90
Fermentation 141–148
Film production 153–154
Free radical polymerization 11

Gas chromatography 126, 130
Gas chromatography-mass spectrometry (GC-MS) 126–130
Glucose 141, 146, 148
Glycerol 142, 148
Glycolic acid 128
Glycolide 78, 80, 96

Headspace-GC-MS 126
Heterofermenters 146
Hexanoic acid 130
Higuchi equation 75, 85, 87
Homofermenters 146
Hydrogels 69, 70, 73
Hydrolysis 50, 58, 70, 71, 78, 81–104
–, massive specimens 118
–, polycaprolactone 122, 123
–, poly(3-hydroxybutyrate) 124
–, polylactide 118
–, poly(lactide-*co*-dioxepan-2-one) 119
–, poly(lactide-*co*-glycolide) 118, 119
–, semicrystalline polymers 118
Hydrophilic/hydrophobic properties 151, 159
Hydroxybenzoic acid 142
3-Hydroxybutyrate, 3-HB 78
3-Hydroxybutyric acid 130, 131
2-Hydroxyethoxypropanoic acid 128
6-Hydroxyhexanoic acid 129, 130
3-Hydroxyveleric acid 130, 131

Induction period 50, 53
Intermediate degradation products 127, 130
Irradiation 70, 96, 102, 103

Ketene acetals, cyclic 12
– Ketoadipate pathway 142
Kinetic order 53
Kinetics 52, 75, 86

Lactic acid 115, 127–129, 145–147
Lactide (LA) 78, 81, 84, 86, 92, 96, 97, 115, 127, 128

Lactone, coordination ring-opening polymerization 16
Lactones 9, 15, 148–149
Lactoyllactic chromatography 126, 130, 131
Lanthanide alkoxides 52
Lipase catalyzed polycondensation 6, 11
Liquid-liquid extraction 125

Macromers 57
Macromolecular architecture 55
MALDI-TOF 123, 130
Microorganisms 121
Microsphere 70–74, 80–88, 91, 97–101, 105
Monomer 70, 71, 76, 81, 86, 88, 91, 94, 104
Morphology 70, 87, 88, 98, 123
Multimodal molecular weight distribution 118
Multiple headspace extraction 126

NMR 130, 149
Non-toxicity 76, 81, 84, 93

Oil-in-oil 100
Oil-in-water 87, 98, 99
Oil-in-water-in-oil 87, 100
Oligo(propylene succinate) 148, 151–160
Oligo(propylene-co-cyclohexanedimethylene succinate) 157–158
Oligomers 127, 128, 130, 131

PDXO 78, 96, 97, 105
PEG 81, 82, 90, 101, 148, 151, 157–160
2-Pentenoic acid 131
PHB 78–87, 103
Phosgene 151, 156, 158
Plasticizers 132, 159
Poly(ε-caprolactone) (PCL) 61, 77, 79, 82–87
Poly(1,5-dioxepan-2-one) 51, 55, 78, 96, 97, 105
Poly(3-hydroxybutyrate) 78, 79, 82, 84–87, 103, 117
–, biodegradation 124
Poly(3-hydroxybutyrate-co-3-hydroxyvalerate) 117
Poly(3-hydroxybutyrate-co-3-valerate) 78, 79, 82, 85–87
Poly(acrylates) 69
Poly(alkylene alkanoates) 3
Poly(ester carbonate) 148, 151, 155–160
Poly(ester urethane) 7, 148, 151–155
Poly(ethylene glycol) (PEG) 81, 82, 90, 101, 148, 151, 157–160
Poly(ethylene oxide) 78, 84, 85
Poly(glycolic acid) 69

Subject Index

Poly(glycolide) (PGA) 77, 79–81, 103
Poly(lactic acid) 69, 80, 97
Poly(lactide) (PLLA, PDLLA, PLA) 77–86, 97, 103, 105
Poly(lactide-co-glycolide) (PLGA) 69, 70, 77, 82–86, 92, 103
Poly(orthoesters) 78, 86, 93–96, 104
Poly(vinylpyrrolidone) 68, 101
Polyanhydrides 78, 88–90, 100, 103, 104
Polycaprolactone 117
–, biodegradation 121–123
Polycaprolactone copolymers 117
Polycarbonate 78, 91–93, 103
Polycondensation 4, 78, 80, 88, 95, 150–156
–, lipase catalyzed 6, 11
Polyesters, aliphatic 1, 2
–, functional 29
–, hyperbranched 23
Polyglycolide 59
Polyhydroxyalkanoates 117
Polyhydroxybutyrate (PHB) 147–148
Polylactide 48, 59, 115
–, biodegradation 119–121
– copolymers 115, 116
– oligomers 121
Polymerization, anionic 11, 45
–, carbocationic 15
–, free radical 11
–, ring-opening 3, 7, 115, 117
Polyphosphazenes 78
Polysaccharides 77, 86, 87
Pro-drug 70, 73, 82
Processing 131
–, additives 123, 131–133
Propanediol 142, 148–160
– Propiolactone 14, 15
Protein 73, 77, 83, 85, 91, 100, 101
Pumps 70, 73
Pyrolysis 128

Recycling 123, 132
Relative humidity 102–105
Ring-expansion polymerization 50, 56
Ring-opening polymerization (ROP) 3, 7, 78–81, 84, 88–91, 96, 115, 117, 148–150

Rumen organisms 148

Sample morphology 153
Shellac 2
Sludge, anaerobic 123
Solid phase extraction 125, 126, 130
Solid phase microextraction 126
Solubility 159
Solution polymerization 50
Solvent evaporation 98–101
Spray-drying 101
Stannous octoate 48
Star-shaped polymers 57
Statistical experimental design 153
Sterilization 77, 96, 101–103
Storage 70, 77, 101, 103–105
Succinic acid 130, 141–160
Surface erosion 118, 120, 121, 123
Synthesis 4

Telechelic polymers 57
Tetraoxacyclotetradecane-dione 149
Thermal characterization 153–159
Thermal degradation 131–133
Thermoplastics 56
Tin alkoxides 48, 50, 52
Tin butoxide 52
Toxicity 72–76, 99, 102
Transesterifications 46
–, agents 49
Triblock copolymers 56
Trimethylene carbonate (TMC) 81, 91–93, 104

Vaccine 74, 83, 91, 100
Valeric acid 130

Water absorption capability 160
Water-in-oil-in-oil 100
Water-in-oil-in-water 100

Young's modulus 155

Zero-order 75, 82, 86, 87, 90, 95

Printing (Computer to Film): Saladruck, Berlin
Binding: Stürtz AG, Würzburg

RETURN TO: CHEMISTRY LIBRARY
100 Hildebrand Hall • 642-3753

LOAN PERIOD 1	2	3
4	5 **2 HOUR**	6

~~ALL BOOKS MAY BE RECALLED AFTER 7 DAYS.~~
~~Renewable by telephone.~~

DUE AS STAMPED BELOW.

MAY 24 2003

12/20/07

U.C. BERKELEY
SENT ON ILL

JUL 23 2015

FORM NO. DD 10
3M 3-00

UNIVERSITY OF CALIFORNIA, BERKELEY
Berkeley, California 94720–6000